ON DESPERATE GROUND

ON
DESPERATE
GROUND

THE MARINES AT THE RESERVOIR, THE KOREAN
WAR'S GREATEST BATTLE

Hampton Sides

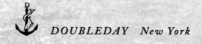

DOUBLEDAY *New York*

www.doubleday.com

DOUBLEDAY and the portrayal of an anchor with a dolphin are registered
trademarks of Penguin Random House LLC.

Book design by Maria Carella
Endpaper maps designed by Jeffrey L. Ward
Title page photograph by David Douglas Duncan, from Photography Collection /
    Harry Ransom Center / The University of Texas at Austin
Jacket image: U.S. troops march south from Koto-ri during the Battle of Chosin
    Reservoir, Korea, December 9, 1950. PhotoQuest / Getty Images
Jacket design by John Fontana

Names: Sides, Hampton, author.
Title: On desperate ground : the Marines at the reservoir, the Korean War's
    greatest battle / Hampton Sides.
Other titles: Marines at the reservoir, the Korean War's greatest battle
Description: First edition. | New York : Doubleday, [2018] | Includes
    bibliographical references and index.
Identifiers: LCCN 2018010543 | ISBN 9780385541152 (hardcover) |
    ISBN 9780385541169 (ebook)
Subjects: LCSH: Korean War, 1950–1953—Campaigns—Korea (North)—
    Changjin Reservoir. | United States. Marine Corps. Marine Division, 1st—
    History. | Changjin Reservoir (Korea)—History, Military—20th century.
Classification: LCC DS918.2.C35 S53 2018 | DDC 951.904/242—dc23
    LC record available at https://lccn.loc.gov/2018010543

MANUFACTURED IN THE UNITED STATES OF AMERICA

10 9 8 7 6 5 4 3 2

First Edition

WALKER AND MACK

*And the steadfast men
of the Korean War Generation*

Sun Tzu says that in battle there are nine kinds of situations, nine kinds of "ground." The final and most distressing type is a situation in which one's army can be saved from destruction only by fighting without delay. It is a place with no shelter, and no possibility of easy retreat. If met by the enemy, an army has no alternative but to surrender or fight its way out of its predicament.

Sun Tzu calls this "desperate ground."

CONTENTS

ON DESPERATE GROUND

MORNING CALM

I n the misting rain, they pressed against the metal skins of their boats and peeked over the gunwales for a glimpse of the shores they were about to attack. Some thirteen thousand men of the First Marine Division, the spearhead of the invasion, had clambered down from the ships on swinging nets of rope and then had crammed themselves into a motley flotilla of craft that now wallowed and bobbed in the channel. Several of the rusty old hulks, having been commandeered from Japanese trawlermen, smelled of sour urine and rotten fish heads. The Marines, many of them green from seasickness, saw the outlines of the charred foothills that rose above the port, and caught the scent of the brackish marshes and the slime of the mudflats. Corsairs, bent-winged like swallows, dove over the city, dropping thousand-pound bombs and sending five-inch rockets deep into hillside nests where the enemy was said to be dug in. Far out at sea, the naval guns rained fire upon the city, damaging tanks of butane that now flared and belched palls of smoke.

On this warm, humid morning of September 15, 1950, the Marines had arrived at their destination halfway around the world, to stun their foe and turn the war around: a surprise amphibious attack, on an immense scale, deep behind the battle lines. Only a few months before, these young men, fresh from their farms and hick towns, had piled into chartered trains and clattered across America to California. Then they climbed aboard transport ships, where many of them did their basic training, learning how to strip and rebuild M1 rifles, drilling on the crowded decks, practicing their marksmanship on floating targets towed from the fantails. They crossed the Pacific and stopped briefly in Japan, then heaved their way through a full-scale typhoon. They rounded the peninsula and moved in convoy up the west coast, through the silted waters of the Yellow Sea.

By the thirteenth of September the ships had begun to concentrate, 261 vessels in all, carrying more than 75,000 men and millions of dollars' worth of war matériel. On the fourteenth, the armada was closing in on its target: the narrow confines of Flying Fish Channel. The channel led to Inchon, an industrial city of a quarter million people, whose strategically vital but treacherous port served the capital, Seoul.

On the morning of the fifteenth, in the seas west of Inchon, the ships ranged along the horizon, a long line of gray bars stitching through the mist. First came the destroyers *Swenson, Mansfield, DeHaven.* Then the high-speed transports *Wantuck, Horace A. Bass, Diachenko,* and the dock landing ship *Fort Marion.* Then the heavy cruisers *Toledo* and *Rochester* and three more destroyers: *Gurke, Henderson,* and *Collett.* Following in their mingled wakes came the British light cruisers *Kenya* and *Jamaica,* the attack transports *Cavalier* and *Henrico.* Farther out at sea lay the fast carriers, from which the Corsairs rose into the skies to make their deadly sorties.

By afternoon, as the Marines drew near the Inchon seawall, enemy rounds zinged across the water surface and mortar shells crashed haphazardly all around. Within a few minutes the Marines would reach the cut-granite ramparts. They would climb over, and they would set foot on the western shores of central Korea.

What was Korea to these young men? Some fate-ravaged place, an inflamed appendix, a wattle of real estate flapping off the face of the mainland. Though some of the units approaching Inchon had spent the past month bolstering United Nations forces at the southern tip of Korea, most of the First Marine Division had no experience with the country. They had little knowledge of Korea's tragic history, no appreciation for how faithfully this proud culture had held on through centuries of turmoil and besiegement. It was a small nation that history had pushed around—a shrimp among whales. The Mongols had had their way with her, the Manchus, the Russians, the Japanese. Now had come the Americans, nervous young men who knew next to nothing about the place, though some had passed around guidebooks filched from hometown libraries.

Korea, they'd learned, was known as the Hermit Kingdom. The Land of the Morning Calm. Its national dish was a hot mess of fermented cabbage. Some Marines had read with astonishment that in parts of Korea, peasants still nourished their crops with night soil and were known to roast the occasional hound for dinner—these were some of the clichés that were passed around. Korea was said to be a dirt-poor country, mountainous, swept by Siberian winds, sultry during summer and breathtakingly cold in winter.

But what did the travel manuals really know? God created war, Twain wrote, so that Americans would learn geography, and the men of the First Marine Division were about to learn a lot about this tough, sorrowful scrap of land. While some would find a deep affinity for it, many would come to hate it, for all the things it was and for all the things it wasn't. But to most of them, more than anything, Korea just seemed a long way from home—a long way to come to fight and bleed and die, in a war that was not officially a war, for a cause that at times was not altogether clear, for an endgame that was anybody's guess.

The Marines had a tradition of being the first to fight, the first to kill, the first to die. They didn't earn their reputation by asking many questions. In a few months, they would face the armies of the most populous nation on earth and would become engaged in one of the more harrowing clashes in the history of warfare. Many would never return home. But for those who survived, Korea would be forever stamped on their psyches, and on their souls. They would never forget what happened here, even if the majority of their countrymen quickly did.

Now the assault vessels angled past the pitted harbor islands of Inchon and made for the bluff face of the seawall, where the breaking surf left a ring of bone-white foam. The bullets whined and smacked in weird patterns upon the surface. A shell screamed overhead, and the Marines crouched a little tighter into the walls of their boats.

BOOK ONE

SEOUL

War is the unfolding of miscalculations.
—BARBARA TUCHMAN

1

THE PROFESSOR

On the Yellow Sea

The amphibious invasion taking place off the jetties and docks of Inchon—an action officially known as Operation Chromite—was among the boldest and most technically complex engagements in modern military annals. The man who conceived the invasion, prevailing over enormous doubts in Washington, was General Douglas MacArthur, and his name would forever be associated with it. But the officer most directly responsible for executing the details of the initial landing, the unsung and largely unknown architect of the Marine assault, was in many ways MacArthur's opposite. He was Oliver Prince Smith, commander of the First Marine Division, one of the great underrated generals in American history.

From the decks of the command ship USS *Mount McKinley,* Smith watched the proceedings as best he could through shifting curtains of smoke. He squinted into his field glasses as the ship heaved in the swells. In the distance, the small landing craft, having butted against the seawall, were in position. In the bow of each vessel, Marines raised a pair of scaling ladders, and the waiting boats treaded the water like enormous bugs with antennae quivering. The leathernecks began to scale the ladders and vanished from view. But the radio reports that came in from the spotter planes were all positive. The Marines were moving into the city now, swarming over the causeways and saltpans, already seizing industrial complexes and other installations along the ruined harbor.

Smith's division was a formidable fighting force. Though it had been hastily thrown together at Camp Pendleton, California, most of the division's officers and noncoms were seasoned warriors, men of the Old Breed who had served in the bitter World War II battles of the Pacific—at places like Guadalcanal and Okinawa. "They could load jeeps with their decorations," said one account, and would "need a truck to carry off their Purple Hearts." Battalion for battalion, the First Marine Division was as fierce and as disciplined as anything the United States had to offer. "It was the strongest division in the world," boasted one Marine captain who served under Smith. "I thought of it as a Doberman, a dangerous hound straining at the leash, wanting nothing more than to sink its fangs into the master's enemy."

Smith was enormously relieved by the progress of the invasion. The resistance was proving tepid—either the North Korean defenders had been caught off guard or they were overwhelmed by the intensity of the firestorm. But Smith stopped short of celebrating. He was superstitious of good fortune. As an assistant division commander at the World War II battle of Peleliu, he had witnessed a senseless loss of life—the result of intelligence failures and strategic mistakes not of his making. Scarred by the events at Peleliu, he tended to proceed with a thoroughgoing sense of caution. In his own experience, it was overconfidence, more than any other single factor, that caused men to die.

He felt a little uneasy that MacArthur had invited a gaggle of journalists aboard the *Mount McKinley* to follow the invasion. It struck Smith as unseemly to have assembled so many members of the press to capture what MacArthur's people confidently expected would be his moment of triumph. "This is a public relations war," Smith wrote in disdain. "We are overrun with onlookers."

Smith had spent a lot of time with MacArthur aboard the *Mount McKinley*. During the past few days, as part of the larger convoy, they had steamed over from Japan in this 460-foot floating command center, a plush flagship tricked out with all manner of radar, radiotelephones, and other advanced communications equipment. Smith found the supreme commander initially impressive, occasionally entertaining, but ultimately insufferable. MacArthur, he said, is "a born actor" who "puts a lot of drama into his conversation." He "has to his credit many outstanding accomplishments," Smith allowed. "However, the pomposity of his pronouncements is a little wearing."

≡

Major General Oliver Prince Smith, fifty-six years old, was a cerebral, soft-spoken man whose habits seemed atypical of a gung-ho Marine. There was no bluster in his demeanor. A Berkeley graduate who incessantly smoked a pipe, he had a reputation in the Corps as an intellectual. People called him "the Professor." One Marine historian described him as an "ascetic thinker and teacher." He was fluent in French, drank sparingly, read the classics, and never cursed. An expert gardener, he cultivated roses in his spare time. He was an incessant notetaker; he kept a small green notebook on his person and wrote in a cryptic shorthand. Reed-thin and tall, his sharp-featured face set with piercing blue eyes and topped by a nimbus of prematurely white hair, he spoke deliberately and with precision.

But for all his gentleness and reserve, Smith was tough. As a young man, he had been a winch operator and then a crew foreman at a rough-and-tumble logging camp in the Santa Cruz Mountains of Northern California. He knew hard work, and his hands showed it. He had climbed his way from an impoverished background to become a scholarship student while steering his family through several tragedies—most notably the death of his only sibling, Peggy, who was found raped and murdered in a cabin at Yosemite National Park, the perpetrator unknown.

On the battlefield, Smith believed in ruthless efficiency. One Marine called him "a professional killer, employed in a hard trade: tenacious, cunning, resourceful, cold, cynical, and tough." He took a dim view of those who brought an exaggerated sense of chivalry to war. Engagements were won by systematically destroying the enemy, not through flamboyant acts or the symbolic capturing of ground. At Quantico, he was famous for giving a lecture that analyzed the effectiveness of the bayonet charge during modern wars; after crunching the numbers, he proclaimed it, by and large, a pseudo-heroic waste of energy.

He was keenly aware of the impact of his decisions. During World War II, through his battles across the Pacific, he would tally in his diary the precise number of casualties the day's fighting had brought. It was his nightly ritual. With the zeal of a sharp accountant who understands where every last dollar was spent, Smith demanded a reckoning of war's exact human cost.

He had enlisted in 1917 and had devoted his life to the Corps. The Marine ethos appealed to his sense of order and rectitude. He had bounced across the globe during his long career: Guam, the Mariana Islands, Washington, D.C., Iceland, the American embassy in Paris. He had lived in a bungalow in the jungles of Haiti and in a castle in the Loire Valley. He had commanded every type of unit from platoon on up. Smith was said to be a "school man and a staff man." He had studied at France's prestigious military academy, L'École Supérieure de Guerre—he was the first U.S. Marine ever to do so. Smith, said one account, was "one of those rare men who love to work and who find a natural delight in detail." If he had come to understand warfare from the perspective of the textbook, he had also seen how poorly and how seldom the theories and abstractions of military science obtained in the context of the grime, grit, and chaos of a battlefield. Smith was a "by-the-book" Marine—but he knew when to throw the book away.

His command style was preternaturally calm. His chief of staff when he was serving at Quantico found Smith to be a rare gentleman, a dignified man who seldom raised his voice: "If you think of a forceful person as one who beats his chest and shouts loudly and utters tirades, then Smith was not a very forceful person. It was contrary to his personality to make a fuss about things. But the people who worked for him listened for any expression of opinion that he gave and took it on themselves as a directive."

Frank Lowe, a retired Army general who was serving as President Truman's eyes and ears in Korea, described Smith this way: "He is a very kindly man, always calm and cheerful, even under the greatest strain. He is almost professorial in type and this characteristic is apt to fool you because he is an offensive tiger. His concept is to find the enemy and kill him—with a minimum of casualties. His officers and men idolize him, albeit he is a strict disciplinarian—*Marine* discipline."

Smith also happened to be one of the country's preeminent experts on the tactics and logistics of amphibious warfare. He had practically written the book on the subject. He had taught ship-to-shore landings in classrooms at Quantico and Camp Pendleton, had perfected some of the techniques on the beaches of Peleliu and Okinawa. The Professor was legendary for his seaside drills. Amphibious operations were among the most complicated maneuvers in war, requiring tedious planning, careful choreography, an exquisite sense of timing. They were the Marine signature, the Marine specialty—at the heart of why the Marines existed in the first place.

Marines were supposed to be "Soldiers of the Sea," shock troops sent to hector coastlines and establish beachheads. They were finned creatures who washed up on hostile shores, only to sprout legs. Time and time again in the Pacific during World War II, the Marines had demonstrated their indispensability, and MacArthur, having relied upon them throughout his island-hopping campaigns, had long been impressed by their steady, ready competence. Smith's First Marine Division was the largest, oldest, and most decorated division in the Corps. So when MacArthur decided, against the better judgment of the Joint Chiefs of Staff, to forge ahead with an audacious and incredibly risky scheme to storm the harbor at Inchon, it was obvious whom he needed to draw up the plans.

≡

The concept behind Operation Chromite—bold, sweeping, magisterial—was much in keeping with MacArthur's style. Throughout his long career, he had shown a preference for the grandiose and the unexpected. A martial romantic, MacArthur loved to speak of decisive thrusts, of hammers and anvils and smiting blows. And, in fairness to him, it was terrifyingly clear that something miraculous had to be done to reverse the course of the Korean conflict.

On June 25, 1950, with little warning, North Korean dictator Kim Il Sung had invaded South Korea with his Soviet-trained, Soviet-equipped army. He quickly took Seoul and steamrolled south in hopes of seizing the entire peninsula. General MacArthur, in Tokyo, did not seem worried at first. "I can handle it with one arm tied behind my back," he boasted the day Kim's invasion began. The United Nations Security Council condemned Kim's aggression and resolved that member states should provide military assistance to South Korea. American troops, later bolstered by U.N. soldiers, were thrown into the breach to help shore up the South Korean army. But by the late summer of 1950, the United Nations forces had been driven down into the southeastern tip of the Korean Peninsula, their backs against the sea. Digging in, they established a perimeter around the coastal city of Pusan. From this toehold they had fought valiantly, but they could not survive indefinitely. Kim Il Sung seemed on the verge of victory.

MacArthur's idea now was not to inject more men and matériel into the Pusan perimeter, but rather to secretly land a huge force farther up the peninsula, well behind the battle lines. With that force, he would cut off Kim's supply chain while swiftly capturing Seoul. (He picked Inchon mainly because it was the port nearest the capital.) MacArthur, who fancied he had an intuitive understanding of "the Oriental mind," argued that capturing Korea's largest city, and doing so on September 25, precisely three months after the start of the war, would wreak psychological damage on the enemy. Asians, he believed, were acutely attuned to numerology. Kim's forces would interpret these developments as a crushing sign that the fates were against them.

At a strategic conference in Tokyo on August 23, MacArthur used the full force of his personality and stature to convince an assembly of skeptical admirals and generals. "We must strike hard and deep," MacArthur vowed. It would be an end-run affair that the enemy could not expect. Instead of going in at the foot of Korea, he would enter at its navel. By preserving the element of surprise—by landing en masse at Inchon and seizing Seoul—he would "seal off the entire southern peninsula." Kim would thus be trapped between the U.N.

forces in Seoul and those in Pusan. True to MacArthur's pet analogy, it would be as though the North Korean troops were caught between a hammer and a mighty anvil.

When he saw that the assembled war council still doubted the plan, MacArthur jacked his rhetoric into the loftiest registers. "I can almost hear the ticking of the second hand of destiny," he said, his voice trembling. "We must act now or we will die." He continued: "I realize it is a 5,000-to-1 gamble, but I am used to taking such odds. Operation Chromite will succeed, and it will save 100,000 lives. We shall land at Inchon, and I will crush them."

$$\equiv$$

General Smith was summoned to Japan from Camp Pendleton in late August to meet with MacArthur and begin formulating the details. It was there, in downtown Tokyo, in the monolithic Dai Ichi Insurance Building where MacArthur kept his headquarters, that Smith began to sense what a peculiarly sycophantic environment the supreme commander had created for himself. Douglas MacArthur, it was said, didn't have a staff—he had a *court*. Smith saw this firsthand.

The seventy-year-old five-star general was, at that moment, the most powerful military figure in American history. MacArthur's career had been long and tempestuous, and marked by stunning precocity. He was the youngest superintendent of West Point. He was the youngest chief of staff of the Army. He had become a general in 1918. During World War II, he had presided over the greatest defeat in American history—the fall of Bataan and Corregidor—but he had also presided over the liberation of the Philippines and the Japanese surrender. Now his array of titles stretched plausibility: supreme commander for the Allied powers. Commander in chief of U.S. Army Forces in the Far East. Head of all United Nations troops in Korea. In addition, as the man absolutely in charge of the occupation of Japan, he was the de facto ruler of eighty-three million foreign subjects.

They called him the American Mikado, the American Proconsul, the American Caesar. They called him El Supremo, the Great Panjandrum. He was a man in love with the vertical pronoun, it was said, a man with "a solemn regard for his own divinity." Asia had become his personal domain, and he seemed to know the Far East better than he knew his home country—he had not visited the United States in fourteen years. It was as though he had become the emperor, and a touchy one at that, jealous of his routines and creature comforts, microscopically attentive to the trappings of power and the nuances of publicity. Since the start of the war, MacArthur had been in every sense an absentee general, running his Korea operations from Tokyo. Though he occasionally flew over to the peninsula for a morning photo op or a quick afternoon reconnaissance, he would not spend a single night on Korean soil during the conflict.

MacArthur greeted Smith warmly, ushering him into his sanctum with fulsome praise. The Inchon landing would be decisive, he said, and the war would be over in one month. Not only that, but Inchon would provide existential security for the Marine Corps. After World War II, there had been talk in Washington of radically downgrading the Marines. With the advent of atomic weapons and the ascendancy of the Air Force, the day of the amphibious landing, some suggested, was over. But MacArthur strongly disagreed, and he believed that Inchon would prove his point. "He pulls no punches," Smith wrote to his wife, Esther, in Berkeley. "He feels that the operation will forever assure the Marines of their place in the sun. He apparently thinks a lot of Marines."

For these reasons, Smith was at first positively inclined toward MacArthur and his plan. But after a few days in Tokyo, he decided that much about the supreme commander's world was weird and cultish. MacArthur surrounded himself with yes men, many of whom dated back to his days in the Philippines. He appeared to have insulated himself from facts he found inconvenient or unpalatable. He dwelled in a hermetic universe of his own making.

Because the people who worked for MacArthur seemed to regard him as a deity, the Inchon invasion, having sprung from his head, was thus a providential undertaking that could not be questioned. "With that staff, MacArthur was God," Smith wrote. "It was more than confidence which upheld him. It was a supreme and almost mystical faith that he could not fail."

But in the end, it wasn't so much MacArthur who worried Smith; it was MacArthur's chief of staff, Major General Edward "Ned" Almond. Perhaps more than anyone in the Dai Ichi Building, Almond lionized MacArthur, and in many ways mimicked his behavior. The ruddy-faced Virginian was a notoriously difficult man. "Ned the Dread," he was often called. A whip-cracking officer, Almond had turned in a decidedly lackluster performance in World War II. But MacArthur, seeing qualities in Almond that others had missed, single-handedly rehabilitated his career. Almond was deeply appreciative. "I shall always be grateful," he wrote. "MacArthur is the only one who has ever given me such a chance."

So blind was Almond's loyalty to MacArthur that many critics came to think of him as a toady. ("Ned the Anointed" was another one of his sobriquets.) In addition to making him chief of staff, MacArthur had named Almond the commander of X Corps, a vast and unwieldy amalgam of mostly Army units that would be wading ashore at Inchon. Organizationally, the First Marine Division would be attached to X Corps, so Smith would have to answer to Almond. In effect, Ned the Dread would be his boss.

It was thus with some trepidation that Smith first met with Almond in Tokyo. From the start, things did not go well. Almond kept Smith waiting for an hour and a half and then brusquely summoned him into his office. Though he was only one year older than Smith, Almond repeatedly called him "son."

"My first impression of Almond was not very favorable," admitted Smith. He found the Army general "supercilious." Almond closely interrogated Smith about his command experience, even though it outweighed Almond's. When they began to discuss particulars of the coming invasion, Smith raised questions about the date and location. Smith and Vice Admiral James Doyle of the Navy had both concluded that Inchon, with its intricate channels, hidden shoals, overlooking heights, and other hazards, was not a suitable site and that the landing should instead be made at a spot called Posung-Myon, some twenty miles to the south. Inchon was the wrong place for the operation, Smith insisted—suicidally wrong.

The tidal differential at Inchon was among the most extreme

in the world; it rivaled that of Nova Scotia's Bay of Fundy. Smith, who had taken the trouble to consult the hydrographic studies as well as the lunar calendar, knew that on the target date of September 15, the disparity between ebb and flood tides at Inchon would be about thirty-two feet. Accompanying the extreme tidal range would be constantly shifting currents. This raised perilous complications. A slightly mistimed landing would strand the Marines on the tidal flat for many hours. If the North Koreans had even a rudimentary force guarding the harbor, the leathernecks would become sitting targets on an expansive mud bog.

Then, too, Inchon had no true beach on which to land, only high concrete seawalls that would require specially designed ladders to surmount them. Further, Smith had good reason to believe that the Inchon harbor was sown with Russian mines.

Almond dismissed these concerns. "There is no organized enemy at Inchon," he said. Smith could see that Almond knew nothing about amphibious landings—how to negotiate the tidal fluctuations, how to prioritize the targets and objectives, how to orchestrate the interplay of sea, land, and airpower to minimize friendly fire and civilian casualties. The existing scheme was poorly conceived, Smith thought. Almond was "fantastically unrealistic" and ignorant of "formidable physical difficulties." And it was also obvious that Almond didn't understand Marines.

It would be a simple operation, Almond assured Smith, and "purely mechanical." The date and location were fixed. Counterintuitively, MacArthur had maintained that Inchon's profound unsuitability as a landing spot was precisely what made it the best place to land: The North Koreans would never expect the Americans to show up there. Almond advised Smith to get to work on the minutiae. Then he dispatched the Marine general from his office.

Here was the beginning of a personal feud that, according to one Marine historian, "would become the stuff of legends." These two men could not have been more dissimilar. If Almond found Smith overly cautious, Smith viewed Almond as rash to the point of being cavalier with the lives of his men. Not that Almond didn't appreciate the personal cost of war—his own son and son-in-law had both died in combat during World War II—but the X Corps commander was very much in the George Patton school of aggressive martial maneuvers. At the root of Smith and Almond's mutual dislike was something larger than a personality conflict, larger than the inevitable differences arising from the rivalry between the Army and the Marines. What was emerging was a clash of command styles and methodologies—opposing views on what war was about, how it should be fought, what its goals should be. Yet these two men were going into Inchon together.

And Inchon it would be, despite the glaring drawbacks of the site. It would rest upon Smith and Doyle, upon the Marines and the Navy, upon their teams of planners and engineers, to make MacArthur's concept a reality—so that Almond's X Corps could safely come ashore. It would be, as Smith put it, "an administrative maelstrom." Smith and Doyle had only two weeks to draw up the blueprint.

≡

Now, on the afternoon of September 15, MacArthur's masterstroke seemed to be working brilliantly, just as he had predicted it would. Smith's Marines had safely reached the city of Inchon, with a minimum of casualties. By day's end, Smith's battalion commanders were reporting twenty-one American dead and 174 wounded: minuscule losses for such a big operation. The last of the landing craft, having been caught by the retreating tide, squatted on the tidal flats like beached whales, and would remain there for the night.

The journalists aboard the *Mount McKinley* tried to capture the scene. "Now and then puffs of smoke came softly from the muzzles of the naval guns, waving lazily like huge indolent fingers in their turrets," wrote Reginald Thompson, a British correspondent. "Overhead, the shells made their invisible sibilant flutter in the sky . . . [and] clouds of dust hid the world under a yellow pall."

"The quake and roar of the rocket ships was almost unendurable," wrote Marguerite Higgins, a celebrated reporter with the *New York Herald Tribune* and one of America's first female combat correspondents. "It looked as though the whole city was burning." Higgins could see "the crimson haze of the flaming docks," could hear "the authoritative rattle of machine guns" as "wave after wave of Marines hit the beach [and] blazed a bloody path to the city."

Where were the North Koreans? They seemed to be thoroughly cowed. Their redoubts were destroyed, and the wooded hills they once occupied had been so completely incinerated that they appeared to have been shaved clean. The way was clear for many tens of thousands of U.S. troops to come ashore beginning the next day. The bold gamble of Inchon was proving to be a huge success. General Frank Lowe, Truman's liaison, who had watched the drama from aboard the *Mount McKinley,* said MacArthur had pulled a "white rabbit out of a hat. . . . I have witnessed a miracle."

MacArthur, watching the spectacle unfold while sitting in a swivel chair on the bridge, was ecstatic. He wore sunglasses and a leather jacket and struck what one Navy observer called a "Napoleonic pose." Inchon, the supreme commander crowed, was "the happiest moment of my life." He could find no fault with the operation. "Our losses are light," he noted, and "the entire command has distinguished itself." He added: "Never have the Navy and Marines shone more brightly."

The Professor beamed with pride, too, but he kept it inside. "The reason it looked simple," Smith later boasted, "was that professionals were doing it."

2
TRAITOR'S HOUSE

Seoul

Twenty miles to the northeast, the citizens of Seoul waited anxiously, bracing themselves for the coming Americans. Seoul was a city of nearly two million war-weary people, the fifth largest in Asia, a crazy quilt of old neighborhoods whose tile-roofed houses climbed over the sawtooth ridges and lush green knobs. Looped like a sash across the city's belly was the broad Han River, its war-damaged bridges shrouded in haze. In the September sultriness, the city lay inert. Trolley cars stood unbudging for want of electricity, empty warehouses sweltered, the footsteps of the few marketgoers reverberated off bare concrete. The streets were desolate but for a handful of trundling rickshaws—merchants hurrying homeward. For the most part, people kept to their houses, clung to their families. They knew what was happening; they'd been through this before. First would come the trembling artillery, then the clink of tank treads on pavement, then the bombs and blades and blood. Hard choices would have to be made in an instant, decisions of allegiance, decisions of life and death. That the citizens of Seoul were used to heartbreak did not make heartbreak any easier.

On the edge of the city, in a hilly neighborhood called Buk Ahyeon Dong, in a district known as the West Gate, a two-story Japanese-style house stood curtained and hushed. Its denizens hardly stirred within. A small child could occasionally be seen in a window, nose pressed against glass. The North Korean authorities had labeled this a "traitor's house." No one was supposed to live here anymore, at least not officially. But, in fact, six children called this place home—

six sibling orphans of war, including twin boys who were less than a year old.

Looking after them was their cousin, Lee Bae-suk, a medical student. He was a boyish young man of twenty, with a delicate voice and a frank face lightly scruffed with beard stubble. In those anxious days, as the Americans approached, Lee played parent as best he could, padding through the darkened rooms, shushing his little cousins, fixing meager meals of noodles or rice. It was a fine house, or had been once—a cheerful place lined with books and classical records. Musical instruments were lying about, and a well-tuned piano was set against a wall.

Lee never ventured outside, for he knew that if he did, there was a good chance he would be seen, maybe captured, maybe killed. One never could be sure who might be watching and where their fickle sympathies might lie. Neighbors spied on neighbors. Friends became accusers. Families turned against themselves.

His cousins knew the procedure: When a noise stirred beyond the fence, when a visitor knocked, Lee would slip upstairs and bury himself in the jumbled contents of a bedroom closet. The older children would shut the door and scoot a large armoire against it. Inside, he would wait in nervous silence until the trouble passed.

≡

Lee's fear of the North Koreans was paradoxical, in a way, for he himself hailed from North Korea. He was born in an industrial city far to the northeast and had been raised there, the oldest of nine children—only to escape the nascent Communist country in 1946, at the age of seventeen. Eluding border guards, he had made a weeklong trek over mountain trails that led toward the South. Penniless and half-starved, Lee had crossed the thirty-eighth parallel, then hitched a ride on a coal truck bound for the South Korean capital.

He knew he had relatives who lived somewhere in Seoul, and after a long search he showed up, begrimed with coal dust, at the house in Buk Ahyeon Dong. His aunt and uncle immediately took him in. They were extraordinary people, warm and generous and kind. At the time, they had four young children. They were educated and enjoyed a different life from the circumscribed existence his own family had lived in the North, where his parents were uneducated merchants, descended from peasants.

Lee's uncle, Ahn-seongkyo, was a concert violinist of some note, and a distinguished professor of classical music at the national college. He had been a child prodigy, had studied music in Tokyo, and, as part of a popular quartet, had regularly performed on radio programs broadcast live across South Korea. But it was Lee's aunt, Ok-seon, who was the firebrand of the household. Ok-seon was a teacher and a civic activist, a bold woman unafraid to speak her mind. Ok-seon published a women's newspaper and became a leading organizer of South Korea's newly formed Olympic Committee, which sent the fledgling nation's first contingent of athletes to the 1948 Games in London.

Lee's aunt and uncle became his new parents, and they prodded him to pursue a career in medicine. Soon he was enrolled in premedical school. To make extra money, he worked as a paperboy, delivering his aunt's newspaper on foot through the suburbs of western Seoul. Then he took a night job at a dormitory complex serving the American embassy, where he quickly picked up English.

Lee often thought of his family back in the North. He missed his mother and father, his eight sisters and brothers. He wondered whether his youngest sister, Sun-ja, who was only three at the time he left home, would remember him. He despaired for his family's fate under the new Communist regime and feared that he might never see them again. The border was becoming increasingly militarized, and passing across it was growing ever more perilous.

Some days, a terrible homesickness gnawed at him. Lee was happy in his new life in Seoul, but happy in a way that sometimes made him feel guilty for his good fortune. For four years, he lived here with his aunt and uncle and their growing family—in early 1950, Ok-seon gave birth to twins. Lee thrived in his studies, too. He was well on his way to becoming a doctor, and hopeful for the future.

=

At the end of World War II, the Allied powers had been faced with the question of what to do with the spoils of the Japanese Empire. Korea had been a Japanese colony, and at the suggestion of the young American diplomat Dean Rusk, who reportedly used nothing more than a National Geographic map as a guide, the peninsula was summarily divided at the thirty-eighth parallel. This line was an arbitrary one—the two "nations" shared the same culture, the same history, the same language. Now the peninsula had been carved up into more or less equal halves, with the understanding that the Soviets would temporarily control the North and the Americans would temporarily control the South, until the country's thirty million citizens could become reunited under one independent government. But that reunification never happened. The two separate realms were quickly remade, each in the image of its custodial nation.

In the South, the United States installed a staunch anti-Communist, pro-capitalist, American-educated leader named Syngman Rhee, who held the country together but proved to be a ruthless authoritarian. When he came to power, in 1947, Rhee's government tortured and assassinated opposition leaders and led a military campaign against left-wing insurgents that culminated in the deaths of more than 75,000 people. This "awkward bedfellow of 'democracy,'" as he'd been called, had proven an embarrassment to the United Nations. The United States, growing ever more focused on the global containment of Communism, tended to look the other way, but Rhee's police-state tactics concerned the Truman administration enough that it refused to provide the republic with much in the way of heavy arms—thus leaving South Korea vulnerable to attack.

In the North, meanwhile, the Soviets had handpicked Kim Il Sung to lead the fledgling Communist nation. Kim had been a ferocious and wily guerrilla resistance fighter against the Japanese in World War II and had become indoctrinated as a Communist ideologue. He proved a master at broadcasting his own legend, exaggerating his exploits as a fighter; among North Korean peasants, stories circulated about how Kim could render himself invisible during battles— it was even said he could walk on water. Consciously imitating Stalin's model, he developed a cult of personality, erecting gargantuan statues and billboards in his honor and calling himself "Great Leader." Later, his public relations minions would go further, declaring him "the sun of mankind and the greatest man who has ever appeared in the world." Tightening his grip on the reins of power, he systematically imprisoned, exiled, or murdered his political rivals. Kim vowed to unite Korea under one government—*his* government—creating what he called a "happy society" that would eradicate all vestiges of "American imperialists and their stooges."

For several years, the South had fended off border incursions from the North, and vice versa. Vengeance killings, guerrilla attacks, and border skirmishes were routine. By late 1949, a simmering peninsula-wide civil war, rooted largely in settling old scores and punishing those who had collaborated with the Japanese, was coming to a rolling boil.

On June 25, 1950, Kim Il Sung's Soviet-equipped army came rumbling across the border, catching much of Seoul off guard. Lee Bae-suk, working at the American embassy complex, looked out the window one day and saw a tank in the street. Its turret swiveled and its gun seemed to be aiming right at Lee's window. At first he thought it was a South Korean tank, but then he realized: It was Russian-made. The North Koreans were already here.

As Kim's troops tightened their grip on the city, foreign dignitaries fled the country, and the government went into exile. South Korea's anemic army retreated south to take up defensive positions at the tip of the Korean Peninsula. With the Korean People's Army (KPA) seizing power, Seoul was thrown into upheaval. Party officials rounded up suspects to face hastily organized people's courts. Indoctrination meetings were held in which citizens were expected to publicly "criticize" themselves and pledge allegiance to the new state. The party staged military parades that featured billboards of Stalin and Kim. Radio stations, carrying signals beamed from Pyongyang, played party anthems and stirring odes to the Kremlin. Meanwhile, the new occupiers settled ancestral scores and committed atrocities, executing "enemies of the people"—however loosely defined—and dumping their bodies into mass graves. More than three thousand people were killed.

In this nightmarish chaos, Lee did not know what to do. He had no job—the abandoned American embassy had been taken over by the North Koreans. He would have to give up his medical studies, too, because he realized that he couldn't risk being seen on the streets. A young man like Lee, of military age, would immediately arouse suspicion among the police. Any soldier from the North would likely catch Lee's accent and know where he was from. There was no hiding it; his dialect would give him away.

And what was a young North Korean man of sound body and mind doing here, living in Seoul? He should have been fighting. He should have been with his comrades, digging in for the American onslaught. The answer was obvious: He had to be a traitor. An enemy of the people.

So, at his aunt and uncle's insistence, he sequestered himself inside. For nearly three months he hid here, in the house in Buk Ahyeon Dong, in the West Gate.

≡

But Lee wasn't the only one in the household who faced danger. Ok-seon had good reason to be fearful, too. Her career as an outspoken community organizer, newspaper editor, and South Korean patriot put her in particular peril. Over the years, she had angered plenty of influential people; many around Seoul considered her newspaper far too insistent and shrill. Upon seizing the city, party officials quickly shut down her press, as they did all others in Seoul—only official party organs were allowed to exist. One day in July, the military police came knocking on the door, looking for Ok-seon. They escorted her to a nearby station and interrogated her for quite some time. They released her but demanded that she return every morning and "report" to the authorities.

This Ok-seon did, diligently, morning after morning. Then, one day, she went to the station and never came back. That was it. No one saw her again. No information was available. Ok-seon had simply vanished.

Ahn-seongkyo tried to cope as best he could. He did not protest his wife's disappearance, for that, he thought, would put his own life in immediate jeopardy. He had six children to raise, seven if you counted Lee. He had to put up a good front. He held out hope that Ok-seon was alive—languishing in some prison, perhaps, but alive. Ahn-seongkyo understood that his having studied music in Tokyo could be a fatal stroke against him. Among other targets, party operatives were said to be on the hunt for Japanese "collaborators" from the war days.

Ahn-seongkyo continued teaching at the university, and occasionally he gave private music lessons in the house. But one morning he went off to school and didn't come back. He, too, had vanished. Lee would never see his aunt or uncle again.

Lee was on his own, the man of the house, responsible for six children. He could only assume that his aunt and uncle were dead. Fortunately, Ok-seon had an older sister named Shin-kyeon, who lived elsewhere in the city. At considerable personal risk, Shin-kyeon would bring food and come to the house to help Lee tend to the children. One afternoon, while Shin-kyeon was there, a visitor rapped on the gate. Shin-kyeon, sick with dread, clicked the latch. Outside the door stood a North Korean soldier, holding a rifle.

The soldier demanded entry and, once inside, began to search the house, upturning furniture, rummaging through drawers and cupboards. What he was looking for wasn't clear. In the vestibule by the front door, he knelt to study the long row of shoes neatly arranged on the floor—children's shoes, mostly. He snatched up a pair that obviously belonged to a grown man. They were Lee's. "Whose rubber shoes are these?" the soldier barked. "Who is the owner?"

Shin-kyeon maintained a facade of calm. "Those old things? They're my husband's. He's at the market today."

The soldier, with a sniff of suspicion, set the shoes back down and left. Lee, trembling upstairs in the closet, had heard the whole exchange.

≡

Now that the Americans were coming, Lee feared that Seoul would erupt in new paroxysms of violence. The North Koreans, in

their desperation to hold the city, might resort to anything: looting, rape, reprisal killings. When the Americans came, there would
likely be door-to-door fighting. The occupiers, cornered and probably
doomed, would lash out against the civilians, Lee thought, like a tiger
caught in a trap.

In those muggy days of September, every sound on the street
caused him to jump. He worried about his cousins. He worried about
his country. Now and then he could feel distant shudders, could hear
planes throbbing overhead. Lee peered through a tiny gap in the curtains and waited.

3

ACROSS THE HAN

Inchon

The American soldiers of X Corps kept wading into port—twenty thousand men, thirty thousand, sixty. With each high tide, more landing ships nosed into the channel. From their maws came tanks and tractors and rolling artillery pieces, bulldozers and jeeps and endless crates of ammo. The invasion was unstoppable now. The bridgehead flared like a fast-growing spore. Ziggurats of war stuff swelled upon the docks—food and fuel, medicines and generators, radio equipment, rifles and grenades and shells—the steady streams of supplies coming in, said one British journalist, "as fast as the brilliant large-scale organizational genius of the Americans could bring them."

General Oliver Smith had been impatient to get off the *Mount McKinley* ever since the invasion began. The Professor finally caught a ride to the seawall on the evening of September 16 and set up his division headquarters in a well-placed nook of the smoldering port as the men of his First Marine Division snuffed out the last pockets of resistance. In just two days, Inchon had been quelled. The city, said one correspondent, was "a burned-out husk of a place that would, it seemed, never live again." Already, Smith was turning his sights on Seoul, twenty-four miles to the northeast.

General MacArthur, still espousing his numerological theories, remained adamant that Seoul must be captured on September 25. The symbolism of the date was of paramount importance to him—never mind that a battle for the city must first be fought, one that was

likely to involve door-to-door combat and tricky matters of tactics
that would present their own demands and suggest their own timetable. The Professor, from his experiences in the Pacific, had learned
to be skeptical of preordained deadlines in the face of battle. But he
would do his best.

MacArthur, marching ashore for a sightseeing tour of the ruined
port with a retinue of aides and correspondents, met with Smith
in his command center. MacArthur urged him to press forward to
Seoul with all dispatch. This Smith was already doing: He had sent
his Fifth regiment toward the northwest to seize Kimpo Field, the
largest airstrip close to Seoul. He'd sent his First Regiment to capture the main road that led to the Seoul suburb of Yeongdeungpo.
His Marines were already closing in on the approaches to the capital. The skies swarmed with friendly aircraft, and American artillery
was pummeling the city's outskirts. The campaign was going like
clockwork.

However, a few obstacles stood in Smith's way. The enemy still
held a few villages, and some stretches of road had been mined. And
there could be snipers just about anywhere. Then there was the Han.
The tidal river, broad and swift, was one of the great waterways of the
peninsula, with tributaries extending far back into the watersheds of
both North and South Korea. The river tumbled from the western
slopes of the Taebaek Mountains, flattened into rich alluvial plains,
then swirled through a maze of coastal islands before spilling into the
Yellow Sea. At Seoul, the Han was a quarter mile across and was rich
with the mingled smells of ocean and sod, its turbid channel full of
needlefish and sticklebacks, mullets and eels.

Smith would have to cross it somehow. All of the Han's bridges
had been blown, and they were not immediately repairable, so the
engineers would have to devise something on the fly. The amphib-
ian challenges of the Inchon landing were thus almost immediately
replaced by this new challenge: how to get thirteen thousand Marines
expeditiously across a river whose currents were treacherous, whose
best ferrying points were fortified, and whose margins were nothing
but mud bogs.

≡

General Ned Almond, the X Corps commander, was deficient of ideas on how to cross the Han. Almond, who had neglected to order extra bridging materials from Tokyo, had left the details for Smith and his people to figure out. As with the Inchon landing, the Army general seemed to think that crossing the river was another mechanical problem that could easily be solved. Smith's chief operations officer, Colonel Alpha Bowser, later scoffed: "General Almond had a habit of treating the Han River like it had five or six intact bridges across it, and of course it had none."

Almond's primary focus seemed to be on satisfying MacArthur's desire to reach Seoul by the hallowed September 25 deadline. He kept prodding Smith and his staff to move faster. So what if a river stood in the way? "There was no dearth of advice about speed and boldness," Smith later wrote.

At least Almond had stopped calling him "son." However, Smith was starting to see a quality in the X Corps commander that was far more concerning: an impetuosity, a tendency toward snap judgments, a willingness to ignore on-the-ground realities in favor of abstract goals. Those who knew Almond's style from his days in Europe viewed him as a whirling dervish—someone who thrived on adrenaline-soaked chaos. Certainly he had shown no lack of courage; in Italy, where he had fought, he gave little thought to his own safety. But he was an impatient micromanager, constantly on the move and seemingly incapable of delegating. "He was what we referred to as a 'hard charger,'" said Bowser. "A rather vain man in many ways." He was "mercurial and flighty," the operations chief thought, adding: "If he had one glaring fault, I would say it was inconsistency."

Then, too, Almond's temper was legendary. Said one historian: "He could evoke the thunders if crossed." Ned the Dread left a long trail of unease in his wake. It was said that he could "precipitate a crisis on a desert island with nobody else around." Alexander Haig, who was then a close aide to Almond, called him "the most reckless man I have ever known." Almond, it seemed, had only one martial modality: *Attack*. Planning and forethought were not his long suits. To him, war was about sallying forth, gaining ground, planting the flag. "When it paid to be aggressive, Ned was aggressive," one observer wrote. "When it paid to be cautious, Ned was aggressive."

So Smith would have to be aggressive, too. As for crossing the Han, he would need to improvise his own solution. The men themselves would cross the river stuffed inside amtracs—amphibious tractors—the wallowing snub-nosed beasts, equipped with twin Cadillac V-8 engines, that served as the tried-and-true assault vehicles of ship-to-shore warfare. The bigger problem was how to transport large equipment, such as the Marine tanks, which weighed forty-two tons apiece. After testing various prototypes, the engineers designed and constructed a klutzy raft set upon large pneumatic pontoons. The jury-rigged barges, said one journalist, were little more than mattresses of timber, but they worked. Though it would take a while, these creaky craft would ferry the tanks across.

Early on the morning of the twentieth, Smith observed the first crossings from a high hill, field glasses pressed to his eyes. "There was plenty of enemy fire," he wrote, "but our people went on in." Like colonies of determined beavers, his men churned across the river, the wakes of their vehicles leaving a bewildering pattern of *V*'s over the broad gray surface. The North Koreans, entrenched in the green hills that overlooked the far bank, put up a fight, but vanguards of the Marines quickly waded ashore, climbed the heights, and silenced the enemy guns.

The crossing was nearly a complete success, and soon Smith's Marines were pressing toward Seoul. They marched along the highway, or rode in trucks, through the terraced hills, past sorghum fields and paddies of rice ready for harvest, past green gardens and orchards pendulous with fruit. Along the way, they encountered a few pockets of North Korean soldiers, but not many—the enemy was offering little resistance in the countryside. Pyres rose over the ridgeline, in the direction of Seoul, and the air was tinged with yellow smoke and dust. Civilians streamed out to the roadside, some cheering or waving flags, others staring with severe looks of confusion and hope.

The scenes grew more desperate, the sense of panic more acute, as the Marines drew closer to Seoul. The roadside became a chaos of rubble and tangled telegraph wires, with refugees hurrying along the roads. It felt as though a storm were approaching. "Women carried huge bundles on their heads and pushed carts that overflowed with belongings," wrote Joseph Owen, who served with the Seventh Marine Regiment. "Dirty children toddled beside bent grandparents, and the troops tossed candy bars to the kids. The Koreans, old and young, scrambled in the dirt for the candy."

≡

As the Marines began to probe the outlying precincts of the city, the human cost of MacArthur's deadline became more apparent to Smith. He felt the September 25 date was contrived—little more than a political gimmick designed to win headlines. Smith reckoned that his Marines could probably take Seoul by the twenty-fifth, but only by laying waste to large sections of the city, pounding it with artillery, bombing it to cinders. Seoul would be badly scarred, and the civilian death toll could be terrible.

Smith knew that there were other, less destructive ways to take the city. Alpha Bowser insisted that the Marines could capture Seoul "with hardly a brick out of place." They could encircle it, cut the enemy's supply lines, and methodically ferret out the defenders, block by block. But this kind of fighting would take more time than MacArthur was willing to tolerate.

So the big guns were brought forward, and the ritual of "softening up" targets across the city began. This, of course, was but a euphemism for a devastating bombardment that could only strike terror in the hearts of Seoul's residents. General Almond was pleased to note that the enemy would be pounded to pieces. That a city might be razed in the process appeared not to trouble him.

The supreme commander seemed similarly disinclined to dwell on such unpleasantries. MacArthur sensed that he was on the brink of total victory, and nothing could dampen his euphoria. The Inchon operation had been a brilliant success. Everything he had predicted was coming true. He had taken the enemy by surprise. Down at the Pusan perimeter, North Korean units were reportedly dissolving and retreating headlong for the thirty-eighth parallel. The U.N. forces at Pusan, including the Eighth Army, commanded by General Walton Walker, were breaking out of their besiegement. MacArthur could already see the war's end.

He was so pleased with the progress of the invasion that the next morning, September 21, he decided to fly back to his headquarters in Tokyo and leave the advance on Seoul to Almond and Smith. (Since the start of the Inchon assault, MacArthur had been bunking on the *Mount McKinley,* never on shore.) In MacArthur's view, taking Seoul would be little more than a mopping-up operation. Throughout World War II, he had shown a habit of publicly declaring targets to be pacified well in advance of the deed—and his old predilections were still in evidence. He seemed to discount reports that more than thirty thousand North Korean soldiers were in the capital now, digging in for a siege. The city would fall, MacArthur insisted, in a matter of days. The North Korean defenders would simply evaporate.

General Smith was puzzled and a little alarmed by MacArthur's confidence. Smith accompanied MacArthur to Kimpo Field, where the supreme commander surprised everyone by cooking up an impromptu award ceremony on the tarmac. MacArthur, his voice quavering, called Smith "the gallant commander of a gallant division." Then he pinned a Silver Star on the Professor's chest.

The Silver Star was one of the highest and most coveted accolades that a U.S. serviceman could win in combat. It was a tremendous honor, yet Smith was mortified by the gesture. Not only did this ceremony seem premature, but, more to the point, Smith felt he hadn't done anything to deserve it. The Silver Star was intended for those who'd performed heroic acts on the front lines, in the heat of enemy fire. "It is meant for gallantry in action," Smith wrote to his wife in disdain, and "not appropriate for a division commander."

MacArthur didn't care. He loved these sorts of martial rituals— the more florid the better—and he genuinely seemed to think that

Smith had earned it. He shook Smith's hand and then boarded his plane for the comforts of Tokyo.

Smith's misgivings about MacArthur were deepening by the day. He stood at attention on the runway, blinking in the morning light, wearing a thin smile of embarrassment and disgust.

Then he turned his attention to the problem of Seoul.

4

BENEATH THE LIGHTHOUSE

Seoul

Even from the depths of the house in Buk Ahyeon Dong, Lee Bae-suk could sense the noose was tightening. The Americans seemed to be burrowing through his neighborhood, pressing inward, street by street, block by block. Outside, Lee could hear jeeps and trucks grinding through the leafy suburbs of the West Gate. The skies swarmed with fighter planes. He could hear mortar blasts, the natter of machine guns. The windows in the house constantly rattled.

Nearby, beyond a rocky outcropping that was a prominent landmark of the neighborhood, he could hear combat. At times, it sounded like close-in fighting, hand to hand. There were screams and cries, radios squawking, commands shouted into the night. The fight for Seoul was entering its last stages.

Lee and his cousins had almost no food left, and they were surly from having been cooped up for so long. He was constantly scolding the children, reminding them to keep quiet. The house, hot and stifling, had become a prison. But Lee knew they had to stay put a little longer; they had no choice. Now, he sensed, was the most dangerous moment of all, the moment of reckoning. The closer the Americans drew to the core of the city, the more maniacal the KPA would become. The most perilous hour, he thought, would be the one immediately before liberation. But he remained optimistic. Lee recalled an expression he'd heard growing up: *Right beneath the lighthouse is the place where it is darkest.*

In quiet moments, he thought about his family in the North.

He wondered if they were alive. What would the U.S. invasion at Inchon mean for them? After the Americans captured Seoul, would they keep on going, pursuing the North Koreans into their own land? Would the U.N. unify the country under the banner of the South, just as Kim had tried to do under the banner of the North? If so, Lee might get to see his family once again. He rejoiced at the thought, and in his mind he pictured a beautiful homecoming. On the other hand, he feared, his family might get caught and killed in the turmoil, as so many Koreans on both sides had. This land was like a thermometer, the mercury rising and falling, north then south then north again. Each incremental change in temperature, in either direction, meant tragedy for someone somewhere.

A few more days passed. Then, on the afternoon of September 23, a calm descended over the West Gate. The fighting, it seemed, had stopped. Looking out an upstairs window, Lee caught an astonishing sight: a large cluster of troops, gathered on his block. They weren't marching anywhere. They were standing around, puffing cigarettes, talking and laughing. Some vehicles were parked nearby, and sandbags were piled around a machine gun. Lee studied them closer and realized—they were Americans.

Then Lee did something injudicious. He bounded down the stairs, flung open the gate, and ran out to greet the Yanks. There was no telling who in the neighborhood might be watching—quislings, perhaps, spies of the KPA. It was possible, too, that the American soldiers, edgy from battle and thinking him a threat, might shoot him on the spot. But in his excitement, he threw caution aside.

"Welcome!" he cried. "Welcome, Americans! Welcome to Seoul!"

He approached them with his arms upraised and flashed a broad smile. Striking up a conversation, he surprised the Americans with his fluency—the English he had learned while working at the embassy was paying off. These men were from the First Marine Division, he learned, under the command of General Oliver Prince Smith. They offered a cigarette and some chewing gum, and Lee stood talking with them for a while. He was embarrassed by his ashen appearance—it had been nearly three months since he last went outside—but he could not conceal his joy.

The Marines were battle-worn and sunburned, dirty and un-shaven. They said they had come up from Inchon, crossed the Han in boats, and attacked the city from the west. Some of the fighting nearby had been horrific. Beyond the outcropping near his house, Lee glimpsed scores of corpses—North Korean soldiers and U.S. Marines alike. A Navy medical corpsman walked among the dead, searching in vain for wounded comrades.

Lee told the Americans he wanted to join them. He was sick of living in the shadows, like some skittish lizard who made his home under a rock. He yearned to shout his allegiance from the parapets. He wanted to be part of the fight.

He returned to the house with a flush of excitement. But the next morning, when he went out to see the Americans again, his heart sank: The Marines were gone. Overnight, the battle lines had been redrawn. The West Gate was North Korean territory once again. By greeting the Americans with such ardor the day before, Lee feared he had exposed himself—and had possibly endangered his cousins as well. Cursing his own stupidity, he slunk back to the house and bolted the door.

5

THE BATTLE OF THE BARRICADES

Seoul

The way into Seoul grew bloodier with each mile. General Smith's Marines finally began to encounter the resistance they had believed would show itself at Inchon. The North Korean People's Army had erected barricades at nearly every major intersection—piles of burlap bags stuffed with dirt and rice and sand, the crude bulwarks sometimes reinforced with furniture and miscellaneous junk hauled out from houses and tenements. North Korean snipers had positioned themselves on rooftops and balconies, in high buildings, in attics and crawl spaces, in cellars with ground-level windows. The streets, meanwhile, were seeded with Russian-made mines. Every cranny of the city had been booby-trapped.

General MacArthur had assured Smith that taking Seoul would be a cakewalk. But MacArthur wasn't there to see it through—he was seven hundred miles away in Tokyo, attending to his cloistered world, surrounded by his court. He would not bother himself with the messy complexities of the conquest; he would follow it from afar, and return in triumph when the dirty work was done.

Because Smith had a deadline to meet—an artificial one, as far as he was concerned—he had no choice but to pummel the city with artillery. It was a decision he hated to make, but once he'd made it, he was lavish with his ordnance. The howitzers, positioned miles away along the Han, unleashed their wrath. For more than twenty-four hours, Seoul's foundations shivered and shook. "Slowly and inexorably," wrote war correspondent Reginald Thompson, "the last life was squeezed and battered out of the city."

Much of the capital was ablaze. The place was a shambles—windows shattered, sidewalks pocked, buildings blistered and yawning with holes. It was, said one Marine, "literally a town shot to hell." A residue of cinders coated the streets, and chunks of concrete crunched underfoot. Downed telegraph wires lay tangled along the roads. Beasts of burden ran in the streets. Thousands of terrified civilians, not sure where to go, darted this way and that, wailing and coughing through heavy smoke that stank of mass death. "Few people," wrote one correspondent, "have suffered so terrible a liberation."

Through this ruined cityscape the Marines advanced, block by rubbled block. They hunched and flinched as they went, for the "crack of bullets overhead was close and constant and meant for them," said Time-Life photographer David Douglas Duncan. Not knowing the proper place names of Seoul, the Marines invented their own—Nelly's Tit, Slaughterhouse Hill, Blood and Bones Corner. Smith, said one journalist, "deployed his men like ferrets into the mole hills of Seoul's waterfront suburbs." Working their way over the city's escarpments, they sometimes ventured into caves, snuffing out the enemy with flamethrowers.

Then the Marines pushed up Ma Po Boulevard, a broad, straight thoroughfare lined with sycamore trees, the trolley tracks that ran down the roadway's center now damaged. The Pershing tanks rolled over the debris-strewn asphalt, swiveling their turrets, poking into structures, occasionally catching the corner of some building's roof and yanking it off while veering into a side street in search of prey.

After the tanks came the sappers and the infantry, their M1s fixed with bayonets. Eyes red, faces smudged with soot, the men skirted the fires and held up their hands to shield their faces from the heat. They crept through railyards, probing and crouching and probing some more, leapfrogging through sniper fire. They kicked in doors, crawled over fences, clomped through gardens. Occasionally the carcass of some smoldering building would collapse in a heap, sending out showers of sparks.

Urban warfare was not a traditional specialty of the Marines—

the World War II veterans among them were more used to fighting on beaches and in jungles—but they adapted to the nerve-racking work. From barricade to barricade, Smith's men inched forward—it was "a dirty, frustrating fight every yard of the way," said one Marine engineer. They blasted the roadblocks with white phosphorus shells, with mortars and grenades and rockets. Red tracer rounds looped through the smoky air and smacked into the sandbags, then the machine guns opened up. The Marines killed all who resisted—one account spoke of "clots" of enemy corpses. The North Koreans who surrendered were stripped naked and, thus humiliated, marched away in droves.

While the demolitions experts moved forward to locate and disable the buried mines, other teams of men hunted enemy gunmen perched in their myriad hiding places. As the Marines skulked through the back alleys, a Molotov cocktail or a satchel charge might rain down from any window. The tension from these forays "whittled us pretty keen," said one Marine. "If one's own mother had suddenly leapt out in front of us, she would have been cut down immediately, and we all would have probably cheered."

Slowly they worked their way along Ma Po, toward the embassies and the main government offices, toward the train station and the Duk Soo Palace of the ancient rulers. One company of Marines had to storm a Catholic church that was being used by North Korean snipers. The church, these Marines discovered, had been converted into a Communist Party headquarters. Its walls had been stripped clean of religious iconography, and huge posters of Stalin and Kim Il Sung leered from the altar. There were propaganda posters, too, depicting American soldiers slaughtering Korean women and children.

While the battle progressed in and around the church, bullets clanged off the large bell that hung on a wooden beam outside the edifice. When the firing began to subside, four brave Korean civilians climbed the tower and tiptoed onto the beam. They stood "boldly against the sky, swinging the bell," wrote Marguerite Higgins, and it resonated "clearly over the racket of the battle . . . a strange, lovely sound there in the burning city." The bell ringers crawled down, and one of them explained to the American troops, "That was for thank you."

As the Marines pressed into the wrecked heart of Seoul, more and more civilians began to spill into the street. Some seemed elated—"hysterically babbling words unknown and offering gifts unwanted to the embarrassed men," said David Douglas Duncan. Other civilians, wrote Reginald Thompson, had "tears streaming down their withered walnut faces, sobbing the most pitiful thanks." But most eyed the new liberators with what Thompson called a "curious impassivity," the look of a shell-shocked people who had endured incalculable hardships and held a tenuous trust in their good fortune. "It was difficult to ignore these people we had come to save," said Thompson, "for the saving had taken on a bitter and terrible flavor."

≡

Speed was what General Ned Almond most cared about now. He was determined to realize MacArthur's wish of seizing Seoul by the twenty-fifth of September. So Almond was constantly on the move across the battlefield, dropping into command posts by jeep and helicopter, surveying the city from a spotter plane, agitating for quicker and bolder modes of attack. The X Corps commander thought Smith was dallying—as he put it, Smith "wasn't in the speed of mind that I was." He later said of the Professor that he "always had excuses for not performing at the required time the tasks he was requested to do." Almond's operations officer insisted that the Marines "were exasperatingly deliberate" at a time when "rapid maneuver was imperative."

Smith *was* deliberate. By instinct and by training, he was a fastidious planner, and he resented Almond's constant goading. But this was only one factor souring the relationship between the two men. The rift had deepened since the Inchon landing. Smith's initial impressions of Almond had been confirmed: The X Corps general was full of swagger. He strutted into meetings, made pronouncements based on a minimum of intelligence after a minimum of discussion, then strutted out. Alexander Haig, who did admire the general for many of his qualities, had to concede that he was an impossible man to work for. "He had to be experienced to be believed," Haig wrote. Almond was "irascible" and "volcanic," said Haig. "Curtness was his hallmark." He had "frosty blue eyes in which there was a perpetual glint of skepticism . . . he drove subordinates unmercifully." When things went wrong, Ned the Dread seemed incapable of self-critique and was quick to blame his subordinates.

In fact, Almond was already famous for this tendency. During World War II, he had been put in command of the Ninety-second Infantry Division, which was composed entirely of African Americans. A curious belief then obtained within the armed forces that white officers from the South were the best prepared to deal with what were considered the peculiar needs and mannerisms of black soldiers. Almond, a native of Luray, Virginia, and a graduate of the Virginia Military Institute, agreed. "Being from the South," he said, "people think we don't like Negroes. But we understand [their] capabilities." At the same time, he added, "We don't want to sit at the table with them."

Such were the widespread prejudices within the Army that, on the basis of his dubious qualifications, Almond was promoted to the rank of major general and put in command of a division of black soldiers in Italy. When his Ninety-second Infantry performed poorly, Almond blamed his division's shortcomings entirely on race, citing what he called certain "characteristics of the Negro and his habits and inclinations." Blacks were piss-poor soldiers, he insisted—lazy, disorganized, apathetic, and cowardly. "Negro soldiers learn slowly and forget quickly," Almond said in a top-secret report after World War II. The white man, Almond asserted, "is willing to die for patriotic reasons. The Negro is not. No white man wants to be accused of leaving the battle line. The Negro doesn't care."

"He was not a believer in the racial integration of the Army," wrote Alexander Haig, "and thought that those who were, such as myself, were in need of education, or perhaps something stronger, to wake us up to reality." Whatever may have been deficient in the training of the black soldiers under Almond's command, whatever may have gone awry on the battlefield, *he* had nothing to do with it. The problem was that he had been given what he called "defectives." Almond, proving a master of damage control, found his way to promotion. He vigorously recommended to his superiors that the Army should never again employ blacks as combat troops, a position he held firmly until his death in 1979.

Now this same man was in charge of a large, polyglot, fully integrated invasion force—one composed not only of African Americans but also Puerto Ricans, South Koreans, and troops from numerous other nations. His views on the inferiority of nonwhites hadn't budged an inch. Almond's racial biases extended to the adversary, too; he had a habit, not uncommon throughout the American armed forces in Korea, of calling enemy soldiers "gooks," "chinks," and "laundrymen." He didn't think much of Asian troops and made little effort to disguise it. He was convinced that, in the end, they would not stand and fight—a view that sharply contradicted Smith's own experiences with the Japanese defenders at Peleliu and Okinawa.

By contrast, Smith, who'd long held progressive attitudes toward civil rights, had personally overseen the integration of the First Marine Division. Under his watch, the induction and training of black troops had taken place without fuss or fanfare. Smith's reasoning was stark in its simplicity: Once a man entered the ranks of the Corps, race played no role whatsoever. A Marine, he said, was a Marine.

≡

Now, as Smith's men fought their way into Seoul, a development in the American media—something that should have been quite trivial—served to deepen the animus between Almond and Smith. That week, for its blockbuster story on the Inchon invasion, *Time* magazine chose to put Smith on its cover. Almond was miffed by this glowing article on Smith and his leathernecks. The Marines may have spearheaded the invasion, but the operation technically had fallen under the aegis of Almond's X Corps. Almond felt that if *Time* hadn't picked MacArthur—the obvious choice—for its cover, then it should have picked him. The influential newsmagazine's decision to single out Smith as the hero of Inchon only excited interservice jealousies between the Marines and the Army, already a source of tension within the ungainly construct that was X Corps. (Smith, ever modest, took a jaundiced view of the *Time* story, grousing that although the recognition was nice, it caused "probably more trouble than it was worth.")

As his regiments bored into the capital, Smith began to entertain more serious differences with Almond—differences having to do with tactics. The X Corps commander kept suggesting ways for Smith to attack Seoul by splitting his division into various advancing prongs that would divert and then boldly reconverge deep within the city. Not only were such schemes logistically convoluted, Smith felt, but they seemed likely to produce casualties from friendly cross fire: If these maneuvers were not executed with pinpoint precision, his units would effectively be shooting at each other. Smith repeatedly expressed concerns about what he called the "coordination of fires." The Professor was a proponent of keeping a division intact as a cohesive fighting force—with each regiment closely supporting the next, advancing in lockstep. He believed in concentrating his men, not fraying them into multiple independent strands. But when Smith raised these various issues, Almond dismissed them.

Then Smith learned, to his dismay, that Almond had been going behind his back and personally issuing orders to his Marine regimental commanders in the streets of Seoul. This violation of the chain of command incensed the Marine commander—it could only sow confusion on the battlefield. According to one eyewitness, Smith "hit the ceiling." He pulled Almond aside and, with "the fury of a patient man," told him to desist. "If you'll give your orders to me," Smith said, "I'll see that they're carried out."

Almond protested that there must be some misunderstanding and denied that he'd given any such orders to Smith's subordinates.

"My regimental commanders are under that impression," Smith countered tersely.

"Well, then I'll correct that impression," said Almond.

The moment lingered in painful awkwardness, and some who witnessed it thought Smith's truculent tone was tantamount to insubordination. He was well aware that the quick-tempered Almond was famous for summarily removing those who crossed him—as one account put it, he had a "propensity to relieve subordinates who gave him dissatisfaction."

But if he had gone too far, Smith didn't care. Almond, he felt, was a creature of hot impulses and raw prejudices, a political tool overly focused on public relations, and someone whose energies were consumed with pleasing a certain august man back in Tokyo. Worst of all, Smith believed that Almond was being reckless—not only with the lives of the men of the First Marine Division but with the lives of the citizens of Seoul.

$$\equiv$$

When the twenty-fifth of September arrived, the North Korean defenders of Seoul still had not capitulated—the Marines were enmeshed in brutal firefights all over the city. But that did not prevent General Almond from releasing a communiqué officially announcing that the capital had been seized and liberated. It read: "Three months to the day after the North Koreans launched their surprise attack south of the 38th Parallel, the combat troops of X Corps recaptured the capital city of Seoul."

The Marines responded to Almond's pronouncement with snickers of derision. Men were still fighting and dying in the streets of the burning city. But MacArthur was going to have his conquest by the magical date, even if it wasn't true. Sneered one correspondent for the Associated Press, "If the city had been liberated, the remaining North Koreans did not know it."

Then, on the evening of September 25, an incident occurred that crystallized Smith's worst suspicions about Almond. That night, Smith received a deeply confusing order from Almond to pursue large numbers of the "retreating enemy." Smith's regiments saw no evidence of a retreating enemy. On the contrary, that night the North Koreans had been launching stout counterattacks. But Almond demanded that Smith's men disengage from their active battle situation and execute a foolhardy pursuit, through the pitch-black precincts of a foreign city, to chase down thousands of "escaping" North Korean troops who, as far as Marine reconnaissance was able to ascertain, did not exist.

Smith knew the order smelled wrong. It seemed to be based on faulty intelligence fueled by wishful thinking. He believed that

Almond had issued it "on impulse without serious consideration of all implications." In Smith's estimation, the assignment bordered on the insane.

But an order was an order, and Smith knew he had no alternative but to pass it on to his subordinates. He radioed one of his regimental commanders, Lieutenant Colonel Raymond Murray, and gave the order. Murray protested vehemently. "I can't pursue anybody," he yelled over the radio. "I'm having a helluva fight to take what I'm supposed to take right now."

General Smith chose his words carefully. "I understand your problem," he said, "but there is a direct order from X Corps to launch a pursuit." Something in Smith's even tone conveyed the diplomatic subtext: This was an order to be heard, then promptly forgotten.

Murray got the message. "Aye, aye, Sir," he said, and then returned to his fighting, which consumed his regiment's energies for the rest of the night and well into the next day. He would never pursue a "retreating" enemy.

The whole matter was disturbing, but Smith had learned an important lesson: In the future, he was going to have to walk a fine line with General Ned Almond. Whenever he got an order from Almond, he would sift and he would scrutinize. He would have to find a way to respect the X Corps commander while taking his own precautions. The lives of his Marines depended on it.

6

THE SAVIOR OF OUR RACE

Seoul

Early on the morning of September 29, General MacArthur and his wife, Jean, boarded a plane in Tokyo, and a few hours later, at 9:30, they touched down at Kimpo Field. The supreme commander had come to preside over an official ceremony in which he would symbolically hand over the keys of the South Korean government to President Syngman Rhee. It was to be a triumphant day, a day of pomp and circumstance—which is to say, the kind of day Douglas MacArthur loved most.

MacArthur had long been a master of military theater, someone who had a knack for putting himself at the photogenic nexus between the military and the political. He had done it in the Philippines, when he waded ashore at Leyte and announced to the world that "he" had returned. He had done it again aboard the USS *Missouri*, when he accepted the Japanese instrument of surrender from Emperor Hirohito. Today was to be yet another histrionic moment in a long career of histrionic moments, another laurel for the American Caesar.

Technically, Seoul still had not fallen, and it was quite disingenuous for MacArthur to proclaim to the world that it had. There was fighting yet to do, but his people had declared the ground situation stable enough to stage a photo-op. They had cleared the roads leading from the airport. They had doused the worst of the fires and sprayed the ditches with disinfectant. They had filled in craters and potholes, raked away debris, and carted off the snarls of downed telegraph wire. They had removed the posters of Stalin, the statues of Kim, the North Korean flags. Now Seoul was ready to receive her liberator.

MacArthur and his wife, along with President Rhee, slipped into a Chevrolet sedan with a license plate that sported five stars. They roared off, followed by a motorcade of forty jeeps. They crossed the Han with ease—over the past few days, Marine engineers, working around the clock under the direction of Lieutenant Colonel John Partridge, had constructed an excellent pontoon bridge. Along the route toward Seoul, military sentinels stood at attention every twenty-five yards. Hundreds of cheering children waved paper Republic of Korea flags. MacArthur seemed in high spirits, grinning and gesturing to the jubilant crowds.

But as the convoy slipped into the city, his expression turned grave. He hadn't realized how much of Seoul had been utterly ravaged. Block after block after block, the skeletal buildings lay gauzed in smoke. The cavalcade had to swerve to avoid slag heaps of rubble. It was a swatch of Armageddon, a jumble of charred stones and crackling ruins. The absentee general seemed sobered by the sight, yet he gave no hint of awareness, then or later, that this circumference of destruction had anything to do with him or decisions he had made.

The closer the motorcade drew to the center of Seoul, the more MacArthur and his entourage could smell the stench of death. In the final days of the battle, Seoul had descended into internecine madness. Before quitting their positions, the Communist forces had wreaked revenge on the families of known South Korean soldiers, on anyone viewed as pro-American—women, children, and the elderly all fell victim. Thousands had been executed. American troops had uncovered mass graves and found houses piled with the dead.

Finally MacArthur and his entourage reached the Duk Soo Palace, where a ring of MPs, wearing sheeny helmets, immaculate white gloves, and spit-polished boots, directed the flow of traffic. At the capitol building, MacArthur pinned medals on some of his favorite officers—including his most favorite of all, General Ned Almond. MacArthur awarded him the Distinguished Service Cross, citing him for his "fearless example" in exploiting the decisive pincer movement at Inchon.

Then, at the stroke of noon, MacArthur entered the main council chamber of the capitol, a great room whose windows were draped in sumptuous purple velvet. As he eased down the staircase with Rhee on his arm, a hush descended over the hall. Photographers snapped away from the balustrades. The room had a large glass rotunda, its shattered panes hanging loose in shards. With every crack of distant artillery, pieces of glass were dislodged from the dome and smashed on the floor, a jarring reminder that the fight for Seoul was not over.

Oliver Smith stood at attention in his dusty green uniform, watching the proceedings along with a few of his officers. Smith disapproved of the whole affair. He thought it unwise to stage a formal public ceremony like this in the middle of the still-smoking city. Worse, he and his Marines had been given the unenviable mission of providing security for the event. Smith's men had had to cordon off and patrol large swaths of the city around the palace, and they were personally responsible for the safety of MacArthur and his entourage. It was a stressful assignment: One sniper in one hiding spot, one well-placed bomb, could cause an international tragedy for which Smith would be held accountable. In his view, this unnecessary pageant had siphoned off large numbers of Marines who should have been fighting.

The supreme commander approached the podium. With his bare head bowed in a pious expression, he recited the Lord's Prayer. Then, in a voice tremulous with emotion, he said, "By the grace of a merciful Providence, our forces fighting under the standard of that greatest hope and inspiration of mankind, the United Nations, have liberated this ancient capital city of Korea." Tears coursed down his face as he turned to Syngman Rhee. "Mister President," he said, "my officers and I will now resume our military duties and leave you and your government to the discharge of the civil responsibility."

Artillery thundered again, and more shards of glass dropped from the ceiling, causing one of Smith's officers to remark, only half in jest, that it was more dangerous here than at the front.

Rhee, the seventy-five-year-old strongman, shuffled to the dais and clasped MacArthur's hand. The president had snow-white hair and wore a crisp gray suit. He had lived much of his life on the run or in exile and had once been horribly tortured by Japanese officers, his fingers burned and smashed. Having studied at Harvard and Princeton, he spoke perfect English, and he considered himself a close friend of MacArthur's. "We admire you," he began, turning his gaze to the supreme commander. "We love you as the savior of our race. How can I ever explain to you my own undying gratitude and that of the Korean people?"

Rhee then looked out over the crowd and voiced a hope that ran deep within the South Korean people. "Let the sons of our sons look backward to this day," he said, "and remember it as the beginning of unity, understanding, and forgiveness. May it never be remembered as a day of oppression and revenge."

As Rhee spoke, the audience could hear the crackle of gunfire not far away, and periodically the fetor of decaying flesh drifted in through an open window. But to Smith's relief, the ceremony ended without incident. MacArthur went back to his motor pool and headed straightaway for the airport. A few hours later he was in Tokyo again, in his digs at the U.S. embassy. General Smith, meanwhile, returned to the business at hand: rooting out the last of the North Korean defenders and liberating Seoul—not symbolically but in fact.

=

Already, MacArthur sniffed a bigger prize. He was no longer satisfied with merely liberating the capital and restoring South Korea to her pre-war borders. The success of the Inchon-Seoul campaign had been so dramatic, the reversal of the war's fortunes so complete, that the supreme commander was casting his ambitions farther north. He knew that his forces could easily cross the thirty-eighth parallel and destroy the last remnants of Kim's retreating army. But why stop there? Why not seize Pyongyang? Why not drive all the way to the Yalu River, North Korea's border with China, and unite the entire peninsula?

What a triumph this would be, what a blow against Communism, against Stalin, against totalitarian regimes everywhere. If MacArthur could pull it off, it would be the crowning moment of his career.

In Washington, the Joint Chiefs of Staff had begun to perceive the same glimmering possibilities. So had President Truman and his secretary of state, Dean Acheson. With the easy victory at Inchon, the mood had quickly turned from gloom to exhilaration. A sense of euphoria filled the halls of government, and few cared to puncture the mood by raising niggling doubts. Nearly everyone could see it: The fruit was ripe for the taking. The Americans had Kim on the run. MacArthur had the firepower and the momentum. The thirty-eighth parallel was an artificial border anyway, a figment of the cartographers. Making it a geopolitical boundary had been one of the "greatest tragedies of contemporary history," MacArthur had said a few years earlier. "The barrier must and will be torn down." Now that it was within his reach, why shouldn't he keep going?

So the stakes had grown, the mission had crept. If it was hubris, it was a strain of hubris in which everyone, not only MacArthur, had acquiesced. "It would have taken a superhuman effort to say no," State Department special envoy Averell Harriman later said. "Psychologically it was almost impossible to not go ahead and complete the job."

By this point, MacArthur seemed virtually unstoppable anyway. After the electric success of Operation Chromite, his stock had never been higher. Few dared to question his logic or his acumen. If MacArthur had ordered his troops to walk on water, said one prominent Army general, "there might have been someone ready to give it a try." Secretary Acheson had taken to calling him "the sorcerer of Inchon." After the astonishing turnabout of the past few weeks, said Acheson, "there's no stopping MacArthur now."

Upon his return to Tokyo, MacArthur found an urgent communication, marked FOR YOUR EYES ONLY, from George Marshall. The secretary of defense essentially had given the supreme commander carte blanche to cross the parallel and forge ahead as far and as fast as he judged necessary. In so many words, MacArthur was being issued a hall pass to go wherever he pleased. "We want you to feel unhampered tactically and strategically to proceed north of the 38th parallel," Secretary Marshall assured him. The document had been approved by President Truman.

By the next day, September 30, MacArthur already seemed to be relishing his new elbow room. "I regard all of North Korea open for our military operations," he cabled Washington. A vicious air campaign began over North Korea that would hardly let up for three years—nearly every city and town, nearly every piece of infrastructure, would be destroyed. By the war's end, millions of civilians would be killed. The stated policy, wrote Assistant Secretary of State Dean Rusk, was to bomb "everything that moved in North Korea, every brick standing on top of another."

As this wholesale devastation proceeded from the air, MacArthur formally demanded the surrender of North Korea. "The complete destruction of your armed forces and war-making potential is now inevitable," he wrote Kim Il Sung. "I call upon you and the forces under your command, in whatever part of Korea situated, forthwith to lay down your arms." Kim gave no reply.

On October 1, a South Korean battalion became the first of the U.N. forces to march across the parallel. Within a few days, large numbers of American troops would follow them. As the military caravans hastened north toward Pyongyang, MacArthur continued to enjoy the full blessing of Washington. The Truman administration issued only one caveat: MacArthur must remain vigilant to any indication that Red China or the Soviet Union might enter the war. At the first sign of their involvement, MacArthur was to halt his advance.

≡

Within a few days, Smith's Marines had vanquished the last remaining North Korean defenders of Seoul. Finally, the capital had been pacified. His men drove north and east of the city and assumed a blocking position in the mountains on the road to Uijeongbu, where their mission was to intercept and destroy any of Kim's forces that might be flooding north toward the border.

Smith had established a temporary headquarters in a decrepit medical building on the edge of Seoul. Its rooms still had the pungent odor of carbolic acid. One of the chambers featured what appeared to be a mortuary slab—Smith had it removed because he found it "too depressing." He and his officers began to set up housekeeping. He got his first bath in a long while, using a bucket and a canteen cup. The general, a lifelong green thumb, was pleased to find a garden out back. Its beds were tangled in weeds, but he was able to gather a fresh bouquet for the table each day.

The plan coming from General MacArthur was that X Corps— including the First Marine Division—would be heading north in a few weeks. But they would not proceed over land. Instead they would return to Inchon, load onto ships, and cruise around the peninsula, sailing far up the coast to a place called Wonsan, a major North Korean port. There they would make yet another amphibious landing and work their way into the rugged mountains of eastern North Korea, marching in the direction of the Yalu. Meanwhile, the Eighth Army, under General Walton Walker, having broken out of its trap at Pusan, would seize the North Korean capital of Pyongyang and then march up the west side of the peninsula. This split-command master strategy seemed dazzlingly complex—with two huge forces, separated by a nearly impassable mountain range, simultaneously working their way north—but dazzling complexity was what MacArthur wanted.

On October 11, Smith returned to Inchon and took up residence in the commander's cabin of the *Mount McKinley*, where he spent his days sketching out the plans for the Wonsan landing. Unit by unit, his Marines trickled in from the countryside. Some found the time to experience the backstreets of Inchon—they tended to frequent "hole-in-the-wall drinking establishments," wrote Joe Owen, of the Seventh Marine Regiment, where "available females" were plentiful but the whiskey was "close to lethal."

General Smith, at work in his floating office, didn't like the way the larger mission in Korea seemed to have changed. He was suspicious of the ad hoc nature of the planning, the sense of drift and overreach that pervaded General Almond's thinking. They were headed north, but no one seemed to know how far. The ultimate objective was not transparent, nor was the timetable. "As to how long we will stay in Korea, I do not know," he later wrote his wife, Esther. "I thought we would have been out of here by now. If we had not gone much north of the 38th Parallel this would have been true."

He also noted how cool the nights had become. He could feel the breath of autumn, a pronounced nip in the air. Beneath his combat jacket he wore a sweater and a wool undershirt. "I hope," he told Esther, "we do not have to operate in this country in the winter."

7

GOD'S RIGHT-HAND MAN

Wake Island, the Western Pacific

Just before sunrise, beyond the rustling palm fronds, a plane rumbled on the horizon. The sound of its Pratt & Whitney engines grew louder, competing with the smash of the breakers that curled along the white coral beach. It was six o'clock in the morning, and the day was already sultry. This tiny Micronesian atoll, in the middle of the ocean, slightly west of the international date line, was home to an array of typhoon-battered Quonset huts built by Pan American Airways. Rusted relics from World War II lay half-submerged in the surf. Over the centuries, castaways had pitched up on these lonely shores, shipwrecked conquistadors and whalers and merchants, and, for a time in the early 1900s, the Japanese had come in search of guano and the feathers of exotic birds like the sooty tern, the masked booby, and the black-footed albatross. If, through most of its history, Wake had been uninhabited, the island boasted at least one modern virtue: It was the only piece of land for a thousand miles in any direction that could accommodate an airstrip, and so it had become the mid-oceanic refueling station of the far-flung American empire.

On this morning—Sunday, October 15, 1950—Douglas MacArthur stood on the crushed-coral margins of the tarmac and waited for the approaching plane to land. The general wore a gold-braided Army field cap and an unbuttoned khaki shirt with five stars fixed to each side of its collar.

MacArthur didn't like being called away from his busy headquarters, not for one minute. "Only God or the Government of the United States can keep me from the fulfillment of my mission," he once told a reporter. But here he stood, two thousand miles from Tokyo.

The approaching plane, a four-propeller Douglas DC-6 that shone bright silver and blue, soared over the island's inner lagoon. The general knew that it was the *Independence*, the official aircraft of Harry S. Truman. MacArthur had never met the president before. He wondered why Truman had chosen, at much expense and inconvenience, to voyage nearly seven thousand miles from Washington—a third of the earth's circumference—to see him on this fragile reef. Truman wouldn't fly all this way on a lark. Had he come to reprimand MacArthur? Was there a critical development afoot? Some secret plan or weapon to be unveiled?

MacArthur didn't know. And for a man who'd built his public persona on a mystique of omniscience, not knowing was the worst kind of torture. Truman's people had offered no agenda beforehand, so the general could only guess what larger intrigues might be at play.

No doubt Korea would be foremost in the discussions. Yet this was puzzling to MacArthur, too. For what was there to talk about? The war in Korea was all but over—the general was sure of it. With Kim Il Sung's army unraveling, MacArthur would keep driving for the Yalu. The Korean Peninsula would be united under a free government that would soon hold elections. An anticipation of triumph pulsed through the high command, in Tokyo and in Washington alike.

It occurred to MacArthur, though he preferred to dismiss it as too obvious, that political calculations might figure into the meeting's rationale. He knew that midterm elections would be held across the United States in a few weeks. He also knew that Truman was bringing a clutch of journalists and photographers with him to Wake. The general wondered: Could this be a political stunt, a gimmick, to allow the president to bask in MacArthur's battlefield glory?

Election publicity was one thing, but MacArthur thought he detected a more sinister subtext as well. He feared this might be a trap. So did his aides. Major General Courtney Whitney, his closest confidant and adviser, waited beside MacArthur on the runway. As the president's plane circled over the island and began its descent, Whitney sensed what he would later call a "sly political ambush."

≡

The *Independence* touched down at 6:30 and taxied to a stop. The morning sun was just then peeking out of the Pacific. President Truman rose from his cabin, radiant and alert, wearing a double-breasted navy suit and a Stetson fedora. Though he seemed in a chipper mood, he was anxious. "I've a whale of a job before me," he had written a friend while en route. "Have to talk to God's right-hand man tomorrow."

The bespectacled Missourian was sixty-six years old and silver-haired, but there was still a snap to his movements as he emerged from the plane. An artillery captain in World War I, Truman had launched his unlikely political career after several years running a men's clothing store in Kansas City. He was MacArthur's direct opposite in many respects, but especially in this one: Throughout his career, people had consistently underestimated him. *Time* had once belittled him as a "queer accident of democracy," but he had a frontier toughness that many people missed. Underneath the brass of his personality, it was said, there was steel.

MacArthur waited at the foot of the stairs. According to protocol, he was supposed to salute his commander in chief, but he only gripped Truman's hand. "Mr. President!" he called out.

"How are you, General?" Truman said. "I'm glad you're here. I've been waiting a long time to meet you."

"I hope it won't be so long next time, Mr. President."

The two men paused on the tarmac, posing for the photographers who were engulfing them. It was, said Truman, a "picture orgy." Their advisers and aides clustered around. General Omar Bradley, the chairman of the Joint Chiefs of Staff, was there. So were Dean Rusk, Averell Harriman, and Secretary of the Army Frank Pace.

Truman and MacArthur marched arm in arm off the runway and slid into the back of a rattletrap Chevrolet sedan that was said to be the best car on the island. A Secret Service agent rode in the front seat with the driver while Truman and MacArthur fell into spirited conversation. The president cut to the subject that was foremost on his mind: What were the chances the Chinese might intervene in Korea? "I have been worried about that," Truman said. It was more a geopolitical, diplomatic, or even psychological question than a military one—who could say what Mao would do?—but the president wanted the general's opinion on the matter.

MacArthur played down Truman's concern. His own intelligence indicated that the Chinese wouldn't dare enter the conflict—and if they did, he was sure his forces would destroy them. He did not think much of Mao's troops. They were nothing more than a band of serfs—subsisting on rice balls and yams, relying on little burp guns and fizzly explosives that usually failed to detonate, an army held together with hemp string and bamboo.

The two men continued the conversation at a Quonset hut, where Truman took a seat in a wicker chair, and MacArthur on a rattan settee. What they talked about is not precisely known. But so far, the two men seemed to be getting along well. Truman found MacArthur "stimulating and interesting." MacArthur thought the president "radiated nothing but courtesy . . . I liked him from the start."

≡

A little after seven thirty, they moved down the beach to a pink cinder-block shack run by the Civil Aeronautics Administration. Secret Service agents posted themselves at the corners. Marine MPs toted carbines. Truman and MacArthur were ushered inside, and the full conference got under way. Nearly two dozen aides and staff members gathered around a long pine table. Truman sat at one end, with MacArthur on his right and Harriman on his left. Fresh pineapple was served, and a light sea breeze soughed through louvered shutters. In an anteroom, an unseen stenographer scribbled shorthand notes.

Truman immediately set an informal tone. "This is no weather for coats," he said. He removed his jacket, and others did the same.

MacArthur eased into the relaxed atmosphere. "Do you mind if I smoke?" he asked, removing a corncob pipe and a pouch of tobacco.

"No," Truman quipped. "I suppose I've had more smoke blown in my face than any other man alive!"

After the laughter subsided, Truman turned to a list of questions he'd prepared in a notebook. First, he wanted to know MacArthur's timetable for the rest of the war.

"Organized resistance will be terminated by Thanksgiving," the supreme commander predicted. "The North Koreans are pursuing a forlorn hope. They are thoroughly whipped. The winter will destroy those we don't." For a moment, MacArthur seemed to rue the coming bloodbath. "It goes against my grain to destroy them," he said. "But they are obstinate. The Oriental values 'face' over life."

The general was so confident in his prognostication that he thought many of his troops would be out of Korea by year's end. Battalions of French, Dutch, and other nationalities that had been requested from the U.N. were no longer needed. He was sure they would never see action.

Then Truman returned to the question he and MacArthur had touched on earlier in the morning. "What will be the attitude of China?" Truman asked. Driving to the Yalu might provoke Mao; it might even spark a world war. "Is there any danger of Chinese interference?"

MacArthur brushed away Truman's question just as he had in private. "We are no longer fearful of their intervention," he replied. "The Chinese have 300,000 men in Manchuria. Only fifty to sixty thousand could be gotten across the Yalu River. China has no air umbrella. There would be the greatest slaughter." The Yalu, he suggested, would run red with Chinese blood.

Astonishingly, neither Truman nor anyone else at the table had a follow-up question to this. Everyone present appeared to agree with MacArthur's analysis. They seemed not to have the slightest concern about the Chinese—or if they did, they didn't raise it. They were dazzled and dazed. The rapture over the war's imminent end, and the magnetic force of MacArthur's delivery, had blunted their thinking.

MacArthur, wreathed in tobacco smoke, nibbled on his pipestem.

"He was the most persuasive fellow I ever heard," said one Truman aide. MacArthur's sanguine view was the one Truman and his men wanted to believe. With victory so close, all other scenarios were inconvenient distractions. A critical moment passed them by—and a fraught and nettlesome question on which many lives might depend was put to rest.

≡

At the Wake Island airstrip, in the tropical glare, Truman offered MacArthur a box of candied plums for his family and then, with cameras clicking, he awarded the general a fourth oak-leaf cluster for his Distinguished Service Medal. He lauded MacArthur for "his indomitable will and his unshakeable faith."

The two men vigorously shook hands. "Goodbye, sir, and happy landing," MacArthur said. "It has been a real honor talking to you."

At 11:35, the *Independence* whirred off for Hawaii and points east. A few minutes later, MacArthur's plane climbed into the air as well, bound for Tokyo. Their parlay had lasted only a few hours. The two men would never meet again.

8

THE TIGER WANTS HUMAN BEINGS

Beijing

At the Zhongnanhai, the former imperial palace in Beijing, Mao Zedong
was in secret deliberations with his
advisers about the Korea situation.
Mao was eager to enter the war. "Another nation is in a crisis," he
reportedly said. "We'd feel bad if we stood idly by." His foreign minister, Zhou Enlai, having recently returned from a series of meetings
with Stalin at his dacha on the Black Sea and gaining his tacit support, concurred.

Mao decided to assign the command of China's armies to Peng
Dehuai, a veteran officer of the civil war and an old comrade from
the days of the Long March. Peng accepted. "The U.S. occupation of
Korea, separated from China by only a river, would threaten Northeast China," he argued. "The U.S. could find a pretext at any time
to launch a war of aggression against China. The tiger wanted to eat
human beings; when it would do so would depend on its appetite. No
concession could stop it." In characterizing the prospect of an American presence on the Yalu, some of the Chinese commanders employed
a hypothetical analogy: The United States would not countenance a
scenario in which the Chinese invaded Mexico and marched right up
to the Rio Grande and the Texas border. That, in reverse, was precisely the situation here.

Peng and Mao agreed on a strategy to entrap the Americans—an
enemy that, they fully realized, had far greater firepower. Peng wrote,
"We would employ the tactic of purposely showing ourselves to be
weak, increasing the arrogance of the enemy, luring him deep into our

areas." Then Peng's far more numerous armies would "sweep into the enemy ranks with the strength of an avalanche" and engage at close quarters. This strategy, Peng thought, would render "the superior firepower of the enemy useless."

A few days later, on October 19, large formations of Chinese troops, the People's Volunteer Army (PVA), secretly crossed the border into North Korea. The word *volunteer* was a calculated political fiction that gave Mao the rhetorical wiggle room to suggest that he was not sending his regular army, and thus had not formally declared war on the United States; this, Mao coyly suggested, was an organic public uprising to defend China's Communist brethren in neighboring North Korea, and he would not stand in the way of it. To further the deception, he ordered his soldiers to strip their uniforms of any insignia that might identify them as being officially Chinese. He seemed to think that American forces might even be stupid enough to mistake his men for North Korean troops. "We all have black hair," he said. "No one can tell the difference." A week later, Mao ordered 200,000 more troops to enter North Korea.

From Mao's perspective, this was a confrontation decades in the making. American imperialism, which the Chairman viewed as merely an extension of the old imperialism of the European colonial powers, had been thwarting China's progress and intervening in her internal affairs for more than a century. He viewed American meddling as a pernicious force, going back as far as the Opium Wars, the Boxer Rebellion, and the disruptive role of American missionaries deep in China's hinterlands. The United States had actively and openly subverted Mao's revolution, supplying arms and assistance to Chiang Kai-shek. Now, from their base in occupied Japan, the Americans appeared to be expanding their sphere of influence throughout Asia. When the defeated Chiang decamped to Taiwan and Mao threatened to attack him there, President Truman had sent the Seventh Fleet to guard the Strait of Taiwan—an action Mao viewed as an affront. Now, by crossing the thirty-eighth parallel and pushing toward China's border, the United States, Mao insisted, was showing an old, consistent pattern.

Yet, privately, Chairman Mao was surprisingly enamored of

much about American culture, and in some senses he wanted China to mimic America's energetic spirit of innovation and technological prowess. Although he reluctantly relied on the Soviet Union in important ways and paid lip service to the ideology of world Communism, Mao also hated, feared, and distrusted Stalin. China's alliance with Russia, seemingly ratified by the signing of the Sino-Soviet Treaty of February 1950, was largely a political sham—the two nations were quite wary of each other and had a delicate relationship. In no way did Mao want China to directly emulate the Soviet system. His vision of Communism took into consideration authentic Chinese thought, Chinese history, and Chinese culture, organically superimposing certain aspects of Marxism-Leninism over a unique national identity. It rankled Mao that neither the United States nor the U.N. would accept Red China's nationhood. On some level he had concluded that perhaps the only way he could get the Americans to take him and his revolution seriously was for China to confront them head-on. It would be a way of validating the People's Republic in the world's eyes. And so it was decided: China would strike.

Besides, Mao reasoned, China had already issued a fair warning to the United States, and to the world, that she would strike. In Beijing two weeks earlier, Chinese prime minister Zhou Enlai had personally told India's ambassador to China, K. M. Panikkar, that if American troops crossed the thirty-eighth parallel, China would certainly intervene. Zhou's ultimatum could not have been stated in more emphatic language. India was one of the few non-Communist countries that had formally recognized Mao's regime as the legitimate government of China, so Panikkar served as a crucially important diplomatic channel. The ambassador immediately cabled Zhou's words to Prime Minister Nehru in New Delhi, who forwarded the statement straight to U.S. and U.N. authorities. But in the end, MacArthur, and Truman's top advisers as well, dismissed the warning as mere Red propaganda, filtered through an unreliable source. China, Mao believed, had been clear in communicating her intentions. But the United States had chosen to ignore the message.

≡

The fifty-six-year-old paramount leader of the newly minted People's Republic of China, having triumphed over Chiang Kai-shek the previous year, was anxious to consolidate his power and flex his muscles on the world stage. Ruthless, paranoid, a devotee of the ancient military philosopher Sun Tzu, and a cunning strategist himself, he was a powerfully charismatic man with odd habits and obsessions. To the alarm of his security police, he was infatuated with the idea of swimming in all of China's major rivers, including the mighty Yangtze, as a way to imbibe the spirit of China. Otherwise, Mao rarely showed himself to the public and conducted much of his state business by his pool, deep inside the palace complex, often wearing a terry-cloth robe and slippers.

When he did go on trips, he did so in secrecy and often brought his own wooden bed with him—on trains, on boats, on airplanes. (He even brought it to Moscow on a trip to confer with Stalin.) Mao refused to pay attention to schedules or conventional expectations about time. He followed no rhythm, circadian or otherwise, and his staff was perpetually perplexed by his erratic habits and spasmodic bursts of energy.

Mao also had an apparently unslakable sexual appetite and believed that orgasms directly halted the aging process. To ward off impotency, he received frequent injections of an extract made from pulverized deer antlers. Although he was married, he had his staff secure him beautiful young women to sleep with—sometimes as many as a dozen liaisons in a single day. He had hideous teeth, rendered dingy brown from chain-smoking and his refusal to practice the most rudimentary oral hygiene—he would only rinse his mouth, once a day, with dark tea. His sour breath was made worse by an infected abscess that he refused to treat. He favored the spicy foods of his native Hunan, an oily diet that often made him constipated, requiring enemas every few days. "A normal bowel movement," said his personal physician, was "cause for celebration among the staff." He also suffered from a mysterious neurological malady that caused hot flashes, profuse sweating, and painful throbbing in his extremities.

Otherwise, Mao was a man of much strength and vitality, though he constantly worried about his health and was certain that his enemies were trying to poison him. A student of ancient Chinese history, an avid mah-jongg player, and something of a poet, he spoke in elaborate and sometimes cryptic metaphors. When trying to persuade Stalin to join him in ejecting the United States from Korea, Mao had warned the Soviet dictator that "if the Americans conquer all of Korea, both China and the Soviet Union will be threatened—like teeth getting chilled through broken lips."

≡

In sending his troops to Manchuria and then into North Korea, Mao had to consider the overwhelming dominance of American air-power as well as the constant overflights of American surveillance. But the simple fact was, Mao's army, often lacking vehicles of any kind, was nearly impossible to spot from the air. It was a foot army, though some units did rely on Mongolian ponies and even, in some cases, Bactrian camels, to haul loads. The Chinese troops slept during the day and moved only at night. Just before dawn, crews would use brooms and pine boughs to smooth over their tracks in the dirt or snow. Many soldiers carried sheets that they pulled over themselves at the first sounds of coming airplanes. They were mostly supplied with rations of precooked rice balls and crackers so they'd rarely have to build fires that might send off smells or give away their position.

During the day, they hid in huts, caves, mine shafts, and railway tunnels. Other soldiers tied themselves to tree trunks so they could inconspicuously catnap during daylight hours. But at night, Mao's armies kept moving south, along ridges, over mountain passes, down the spine of the Korean Peninsula. As they traveled, they were supposed to live off the land as much as possible, subsisting on whatever they could beg or steal from North Korean peasants. God help any living thing that dwelled in the path of this ravening army—deer, oxen, horses, cats, rats, and dogs quickly disappeared. Troops plundered gardens and root cellars in search of anything resembling food—in one account, a Chinese soldier spoke of savoring the supreme delicacy of a raw potato frozen solid; eating it, he said, was "like licking a rock."

Most of Mao's soldiers were powerless and desperately poor young men. They came from the lower echelons of an ancient society that did not particularly value the individual and had traditionally viewed warriors as an expendable class. ("As you do not use good metal for nails," went an old Chinese proverb, "so you do not use good men for soldiers.") If Mao regarded the legions he was sending into Korea as noble patriots, he also seemed to view them as cannon fodder, war trash, little more than slaves to do with as he pleased.

One of the many thousands of soldiers in Mao's army was an eighteen-year-old illiterate peasant from Sichuan Province named Huang Zhi. Huang's background was typical of the war-toughened men who made up the ranks of the Ninth Army Group. His upbringing had been tragic: A famine had separated him from his parents when he was only six, forcing him to forgo the possibility of even a basic education. "Growing up, I had no dreams or ambitions," he said, "other than to keep myself from hunger." After many years spent wandering the countryside, performing servile work for various feudal landlords, Huang joined the army and took up the fight, at the age of fourteen, against the Japanese. Later, during the Chinese Civil War, Huang fought on both sides. The Nationalists, the Communists—it didn't matter to him, as long as he could eat. Constant upheaval had instilled in him a fatalistic sense that if he wished to remain alive, he would need to stay alert and learn to "bend like young bamboo." In October of 1950, Huang's unit, the Twenty-seventh Army, had been in the south, preparing to invade Taiwan, when he and his comrades were abruptly put on trains headed north.

Only when they had arrived in the northeastern province of Antung and begun to cross the Yalu River were they told they would be fighting the forces of the United States. "I didn't have any idea about Americans," Huang said. "I had never met one in my life. But my officers kept telling me never to underestimate the American imperialists. They were invaders; they were our enemies. Personally,

I didn't care about any of this. I was just a soldier. I was there to take orders."

"They said imperialism was on our very doorstep," said Yang Wang-Fu, then a twenty-eight-year-old soldier from Shaanxi Province. "That's what we were told. The imperialist criminals were at our door—they were preparing to break into our house. We were told the Americans would invade China just as soon as they overtook Korea. So we had come to defend the homeland."

Men like Huang Zhi and Yang Wang-Fu had been given precious little to fight with. Mao and his generals were keenly aware of this. They knew the American forces enjoyed every conceivable advantage: better weapons, better transportation, better communications, better logistics. But Mao believed that his armies had something the Americans lacked. *Fighting spirit,* he called it. Mao contended, with something like religious fervor, that his men possessed an innate martial quality that was uniquely potent—a zeal, a lust for battle, powered by an all-suffusing patriotism and the camaraderie of the revolution. The Americans depended on machines, he said, whereas the Chinese depended on people. If the Americans fought with planes and tanks and bombs, his army would rely on such human qualities as surprise and flexibility of movement. Of course, it would also draw on overwhelming numbers. Wrote Mao: "China, though weak, has a large population and plenty of soldiers."

In the end, Mao scoffed at America's supposed superiority—even its atomic weapons were no match for China's fighting spirit. America, he thought, was but a paper tiger. "Weapons are an important factor in war, but not the decisive factor," he said. "It is people, not things, that are decisive." The United States would fight in its way, and he would fight in his. "The enemy can use nuclear bombs, and I can use my hand grenades," Mao reasoned. "I will find the enemy's weakness and chase him all the way. Eventually, I can defeat him."

BOOK TWO

TO THE MOUNTAINS

Those who wage war in mountains should never pass through the defiles without first making themselves masters of the heights.

—MAURICE DE SAXE, *Reveries on the Art of War*

9

MANY, MANY

Wonsan, North Korea

The troopships of X Corps departed Inchon in mid-October and sailed down the coast through the Yellow Sea. The convoy of more than seventy vessels passed Kunsan and Mokpo and rounded the peninsular horn, swerving through a confusion of coastal islands and then turning into the Korea Strait. From the railings, off the port side, the men could see the liberated siege grounds of Pusan, site of so much brutal fighting only a little over a month earlier. Then the transports turned into the stormy Sea of Japan and worked their way up the east coast, past Yeongdeok and Samcheok, past Donghae and Yangyang. Finally they crossed into North Korean waters and steamed for Wonsan, a port city of 75,000 people tucked into a large bay a little more than a hundred miles north of the thirty-eighth parallel.

But as they approached Wonsan, to the men's consternation, the ships turned around and started sailing back down the coast for Pusan. No one seemed to know why. Had their orders changed? Was the war over? Were they going home? Then the ships turned around once again, resuming their northward crawl—only to be followed by yet another turn. The Marines and soldiers of X Corps, crammed into their vessels, didn't understand what was happening.

Eventually the word sifted through the ranks: The North Koreans, working with Russian experts, had mined the waters off Wonsan. Having anticipated that the U.N. forces might land here, they had gone out into the harbor in diverse local craft—barges, junks, tugboats, fishing sampans—and sown the waters with explosives, mostly

Russian-made. The harbor was infested: Thousands of contact mines and magnetic mines bobbed just beneath the surface.

So American minesweepers, along with teams of Navy frogmen, were brought in to clear the approaches to the harbor. More than two dozen of these peculiar vessels went to work, often with helicopters buzzing overhead to serve as spotters. Minesweepers had elaborate wire structures, extending far out from the bows, that were equipped with various floats, depressors, and cutters strong enough to sever the steel cables that often moored mines to the seabed. The sweepers plied the harbor, clearing one long channel at a time, even as North Korean artillery shelled them from shore.

It was tedious but also perilous work: On October 10, two American minesweepers missed their quarry and were blown apart. Twelve men died in the explosions, and dozens more were wounded. A week later, a South Korean minesweeper was also destroyed. The men found one mine—also Russian-made—that had a particularly diabolical design. A dozen ships could pass over it without incident, but the thirteenth ship would cause it to detonate. "It took a curious sort of mind to come up with a notion like that," wrote one Marine, wondering if the number thirteen had a "sinister connotation for Russians as it did in the States."

Given the dangers in the harbor, the X Corps landing obviously would have to be delayed until the sweepers had completed their painstaking task. And so the troopships churned back and forth along the coast—changing direction every twelve hours. The Marines dubbed this endless backtracking the "Sail to Nowhere" and "Operation Yo-Yo." For nearly two weeks, they remained at sea with little to do but watch the dull landforms slide by. As food supplies dwindled, the galleys served mustard sandwiches, glops of fish-head chowder, and other highly dubious fare. Joe Owen, of the Seventh Marine Regiment, called it an "ordeal of misery and sickness, malaise and dreariness. The holds stank of unwashed bodies and sweaty clothes." As one Marine account put it, "Never did time die a harder death."

What made their seaborne imprisonment more difficult to take was their discovery, by radio, that Wonsan had already been pacified. Republic of Korea troops, working their way overland from Seoul, had arrived in Wonsan and quelled all enemy resistance there. The First Marine Air Wing had set up shop at a nearby airfield, and planeload after planeload of men and supplies had safely landed. The zone around Wonsan was deemed so peaceful, in fact, that the entertainer Bob Hope had already dropped in to perform one of his USO comedy routines for the aviators—during the show, he boasted of how he and his dancing girls had beaten the famed leathernecks ashore.

≡

Finally, on October 26, the harbor was declared more or less safe, and the ships bearing the Marines and the other troops of X Corps moved into the inner harbor. It was a cool autumn day, with a dusting of snow on the Taebaek Mountains. Wonsan appeared to be a cheerless industrial city of oil refineries and factories, many of which had been bombed. But it was a strategically important place, a provincial capital, blessed with an excellent natural port. In the distance, to the south, rose Mount Kumgang, a misty crag of granite that for centuries had been a sacred place of pilgrimage for artists and poets. The mountain had a different spirit name for each season. Now it was known as Phung'aksan, "Great Mountain of Colored Leaves." But soon, with the coming of winter, it would be called Kaegolsan, "Stone Bone Mountain."

General Oliver Smith waded ashore and hastily established his command post in Wonsan. The more than twenty thousand men of the First Marine Division (additional Marine units had become attached since leaving Inchon) soon began to stagger onto the beaches as well—and after them came their tanks and personnel carriers, their ambulances and artillery pieces, and their other rolling stock. Apart from the harbor mines, the enemy had put up almost no resistance. What Smith had planned as an amphibious assault was officially demoted to an "administrative landing."

Although the Red Koreans had quit the place, they'd left booby traps everywhere. After coming ashore that day, a pair of Marines marched down the beach, scavenging for driftwood to build a bonfire. In short order they found a pile, but as they started to rummage through the stuff, they triggered a terrific explosion. According to an official report, the two young men were "literally blown to pieces; it was impossible to identify the remains of either and they were buried in a common grave."

Significant swaths of the beach, X Corps also discovered, were laced with buried land mines. But South Korean troops, newly arrived in Wonsan, came up with a brutal solution to the problem: They press-ganged a group of North Korean prisoners and, working in a grid, systematically marched them along the beach. "It was a surrealistic scene," recalled Alexander Haig, who witnessed it, "with men stepping on mines and being blown to bits and the others closing the interval and marching stolidly onward." Eventually, nearly every one of the prisoners had been killed by their own army's explosives.

Despite having left these grim calling cards, the North Koreans were hardly a force to be reckoned with anymore. The battlefield situation was fluid, but so far, the North Korean soldiers whom patrols had encountered out in the provinces seemed disorganized and demoralized. They could be tough fighters when cornered, but they were little more than bands of starving guerrillas at this point, hiding during the day and foraging for food at night as they limped toward the north. An intelligence report from MacArthur's Far East Command in Tokyo was almost boastful in its tone: "Organized resistance on any large scale has ceased," the report asserted. "North Korean military and political headquarters may have fled to Manchuria. The enemy's field units have dissipated to a point of ineffectiveness."

Kim's troops might attempt small-scale delaying actions and scattered ambushes, but X Corps would make quick work of them—General Almond was sure of it. It was going to be a walkover.

When General Smith went to see Almond at the X Corps headquarters in Wonsan, he thought the atmosphere hovered somewhere between giddy anticipation and triumph. Almond seemed to be feeling invincible—he had just been featured, as Smith had several weeks earlier, on the cover of *Time* magazine. As far as Almond was concerned, the war was nearly over. Already he was talking about which units he would be sending home from Korea. But now he wanted Smith to organize his three regiments as quickly as possible and get them out in the field to clear the coastal corridor. Then he wanted Smith to keep going north.

The new plan was for most of Smith's division to hasten forty miles up the coast, then turn toward the northwest, across the plains and into the Taebaek Mountains—the extensive range that forms Korea's spine, the so-called dragon's back. They would move up a narrow road that twisted for more than seventy miles into the highlands, toward a large man-made lake that, according to the old Japanese maps the Americans were working from, was called the Chosin Reservoir. Upon reaching its shores, Smith's men were to keep marching for the Yalu, which was another hundred miles to the north by way of a patchwork of roads. Almond called for maximum speed, much as he had during the battle for Seoul.

From the start, Smith was suspicious of Almond's plan. Among other things, it meant that his division would be strung along a narrow mountain road for nearly one hundred miles, in a long train of men and vehicles. They would move along this single artery, relying on a supply chain that the enemy could sever at any point. In this desolate country, there were no airstrips, no functioning rail lines, no other ways to receive reinforcements or evacuate casualties. They had only one road in, and, should anything happen, only one road out. The more they progressed, the more they would stretch themselves— and the more their survival would depend upon this fragile umbilical cord. Over the millennia, countless battles had been won or lost solely on the question of supply: An army, went the old saying, marches on its stomach. A clear understanding of how to provision one's troops was the sine qua non of any offensive maneuver, the most obvious question that any general would raise before considering an advance. But Almond had given the matter little thought.

As far as Smith was concerned, diffusion of this sort was a cardinal sin for any moving army. It violated most everything he'd learned in military school in France, and it violated what he'd taught his own Marines at Quantico and Pendleton. Generally speaking, the goal, the watchword, was to concentrate, not disperse. A commander never wanted to put himself in a position where his division was spread thin over long distances of hostile country—such a disposition made it virtually impossible for his various units to support and defend one another. Smith told Almond that the First Marine Division would be weakened by dispersion. Almond ignored him.

≡

The forbidding nature of the terrain was the other compounding factor, of course. It was hard to tell by the old topos they'd been given, but the way to the Chosin Reservoir seemed to Smith like classic ambush country: narrow passes, blind curves, bottlenecks everywhere. On the maps, it looked like a nest of tapeworms. He understood that his men would be nudging into the unknown, a mountain fastness where the enemy could lie in hiding just about anyplace. For Smith, certain aspects of the terrain brought to mind the ancient Battle of Lake Trasimene, where Hannibal's Carthaginian forces surprised and devastated the Romans along the shores of an upland lake in central Italy—resulting in the largest ambush in military history.

Smith could certainly work his way up that road, but he wouldn't want to do it in a hurry. The usual technique for moving into mountainous country, he knew, was to throw out flanking patrols to secure the highlands along the route. These flankers, forming a kind of perimeter from above, would leapfrog ahead on the ridges while the main body of men and vehicles advanced along the road below. That way, all the units progressed as an interlocking whole. But maneuvering like this took a lot of time. It would be slow going—much slower than Almond cared for.

Almond discounted this talk as so much priggish pedagogy. Underlying Smith's fretting was the supposition that the Marines might encounter any viable foe at all. It was nothing but wide-open country up there, a land of desolation dotted with a few piddly-shit towns. The Marines probably wouldn't see a soul in those mountains. Smith was worrying about an enemy that didn't exist. If his Marines would move quickly and get to the Yalu, the war would be over, and everyone could go home.

The First Marine Division was only one of three grand prongs Almond planned to launch toward the Yalu. These three columns were to advance as distinct lines—each one physically separate from the others and, given the difficult terrain, each one in a poor position to help the others should any one of them encounter trouble. Progressing along the east coast was the I Corps, an assemblage of South Korean divisions. The next prong to the west was the U.S. Army's Seventh Infantry Division; Almond wanted this unit to pass by yet another hydroelectric impoundment, the Fusen Reservoir, en route to the Yalu. Smith's Marines would be X Corps's westernmost column advancing toward the Manchurian border. Then there was the U.S. Army's Third Infantry Division, which would be held in reserve, behind the other prongs. Finally, Almond planned to create a number of smaller "task forces" that would splinter from the various entities and march in different directions to pursue special errands, large and small.

As it was, X Corps had almost no cohesion as a fighting force: It was an ad hoc ragbag of troops, now more than eighty thousand strong and growing, an administrative monstrosity made up of units from multiple services and multiple countries. To Almond's dismay, X Corps was also composed of a farrago of ethnicities. He actively disliked having so many black troops under his command, and, when possible, he tried to limit their battlefield participation to driving and other menial noncombat tasks. But what most concerned Almond was a unit of Puerto Ricans, the Sixty-fifth Infantry, which, on the basis of no particular evidence, he regarded as thoroughly incompetent. "I don't have much confidence in these colored troops," he said of the islanders.

Almond's piecemeal, multilinear plan for advancing on the Yalu served only to fracture X Corps further. The maps in his headquarters showed confusing Gordian knots and crisscrossing vectors. Almond conceded that the projected battlefield was starting to look a bit complicated—"I got troops scattered all over Korea," he said—but he was confident that his thrusting prongs would achieve their objectives, that out of this panorama of chaos some form of order would prevail.

Smith and his officers were astounded by what they regarded as the haste and heedlessness of Almond's plan—not only the Marine part of the puzzle, but all of it. A childish naïveté permeated Almond's ideas, they thought. He tossed around divisions willy-nilly, as though he were playing a game of jacks. "I questioned his judgment," Smith's operations chief, Colonel Alpha Bowser, later said. "I think General Almond pictured this in his mind's eye as a sweeping victory that was suddenly in his grasp."

At one point, Smith told Almond that with winter around the corner, it seemed imprudent to penetrate too deeply into the mountains. Even if they could reach the Yalu, Smith did not know how X Corps could hold positions across the snowy north country and keep them resupplied through a North Korean winter.

Almond derided this. Smith, he felt, was being a worrywart. "It was abundantly clear to me," wrote Almond, "that what General Smith was complaining about was the fact that his division happened to be the division [being] used to push into the forward area and meet an unknown force." Smith, Almond said, "was overly cautious of executing any order that he ever received."

Almond was flying high now, and those who sat with him in meetings saw a man who had come to view himself, and the forward drive of his legions, as unstoppable. "He was the cock of the roost," said Sam Folsom, a Marine aviation officer who took part in the daily briefings. "His operation was unspoiled by reality. Something in his manner, something beyond rank, told us in no uncertain terms that he was superior—superior to all of us. He was *it*. I don't think he understood what was happening in the field. I don't think he cared to understand."

≡

By now, however, rumors had seeped into Smith's command post that a new enemy had entered the war. At first they were only that: rumors, whispers in orchards and alleyways, furtive looks, premonitions. But then, on October 31, five days after Smith had landed at Wonsan, the first hard report trickled in.

Smith's Seventh Regiment had begun to dispatch reconnaissance patrols to study the road leading up to the Chosin Reservoir. Upon reaching the village of Sudong, about twenty miles inland at the base of the mountains, one of these patrols met a unit of South Korean troops and found them positively rattled. They reported that they had just engaged in a firefight with the Red Chinese—more commonly referred to, among American commanders, as Communist Chinese Forces, or CCF. When asked how large a force they had encountered, the South Koreans would say only that there were "many, many."

But the South Koreans had captured sixteen prisoners and, upon interrogating them, had learned that they were from the 370th Regiment of the 124th Division of the People's Liberation Army (PLA)—Mao's army. They claimed they had crossed the Yalu River in mid-October. These Red soldiers were surprisingly forthcoming with information; they seemed to have nothing to hide. It was almost as though they wanted the U.N. troops to know who they were and where they had come from. They freely indicated that they were part of a much larger Chinese force, numbering in the hundreds of thousands.

When this alarming report and several others like it were sent up the chain of command, Almond's intelligence people reflexively disputed their accuracy. "This information has not been confirmed and is not accepted at this time," a X Corps intelligence memo curtly responded. Almond's headquarters admonished officers in the field to stop conveying the "erroneous impression that CCF units may be engaged."

This strange refusal to believe credible ground intelligence was almost immediately echoed in an official G-2 document sent out from MacArthur's headquarters in Tokyo. "There is no positive evidence," this document insisted, "that Chinese Communist units, as such, have entered Korea."

General Smith was flummoxed by these conflicting signals—those coming from the field versus those coming from Tokyo. In a fog of uncertainty, Smith moved his headquarters north, to an abandoned engineering college on the outskirts of another important industrial city farther up the coast. The city was called Hamhung.

10

KING'S ENVOY TO HAMHUNG

Hamhung, North Korea

n their advance to the North, the soldiers and Marines of X Corps found that they needed raw South Korean manpower of all kinds. They needed cooks and stevedores, they needed runners and fixers, drivers and clerks. Perhaps more than anything, though, they needed interpreters. The language barrier was an exasperating problem—not only the gap between Korean and English but also the one that separated both languages from Japanese. As it turned out, all the maps and logistics manuals the Americans had to work with had been published in Japanese—they were documents dating back to the days of the colonial occupation. The commanders of X Corps recognized that maneuvering over this tortuous land involved a nearly constant filtering, back and forth and back again, among three different tongues.

So recruiters from X Corps had gone hunting among the neighborhoods of Seoul for a fresh crop of trilingual interpreters. This was not an easy task, for nearly all young South Korean men of military age were off fighting somewhere, or they'd already been wounded or killed in combat. Eventually, though, the recruiters found a promising cadre of interpreters, and now they were preparing to send them north.

One of them was a nineteen-year-old medical student from the West Gate of Seoul: Lee Bae-suk.

≡

During the fight for Seoul, Lee had lain low for a few more worrisome days as the North Korean resistance finally withered. When the Americans regained control of the West Gate, he emerged from the house in Buk Ahyeon Dong—this time with confidence. Shin-kyeon, his elder aunt, assumed control of the household and essentially adopted the children, while Lee reported to the X Corps headquarters to offer his services. In the interviews, he had to be circumspect. He stated that he had been born and raised in Seoul. Disclosing that he was from the North, he thought, would automatically disqualify him. Worse, the Americans might suspect that he was a Communist and imprison him as an infiltrator or spy.

Lee made a good impression, and the X Corps recruiters hired him. He would be a translator, assigned to work with American military police at strategic guard posts and chokepoints. With the situation rapidly changing in the North, Army planners kept him waiting in Seoul for several weeks, on standby. Finally, in October, they called him in to discuss his first assignment. X Corps would be sending him far to the northeast, up to a place called Hamhung. It was an important industrial city, pivotal to the fight. The recruiters explained that nearly 100,000 U.N. troops would be pouring through Hamhung, nearby Wonsan, and their environs, en route to the Yalu. General Smith's First Marine Division was already there.

As he heard this, Lee was so overcome with emotion that he could scarcely follow the rest of what the officer had to say. Hamhung was his hometown, the place where he was born and raised. It was where his family lived—if his father and mother and siblings were still alive. Of all the cities in the world, Hamhung was where he most wanted to be. He couldn't tell the American officer this; he had to keep the secret to himself. It took some effort to maintain a straight face, to mask his joy. He had dreamed of Hamhung often, had missed it terribly—the murmurings of the market, the brackish smells of the river, the line of foothills to the west, the ocean breeze at the city's back. Perhaps he'd grown to romanticize the past, but the places of his youth were deeply etched in his memory. He'd left as a boy and had become a young man. He'd been away four years. Now he was going home.

=

Built along the coastal plain, downslope from the Taebaek Mountains, Hamhung had for centuries been a drowsy provincial capital, its uneventful history punctuated by a few storybook moments of intrigue. In 1400, Yi Sung-ke, founder of what became known as the Yi dynasty, took refuge in Hamhung after he was ousted in a palace coup engineered by his own son. When the upstart emperor sent a succession of diplomats from Seoul to reconcile with his father, Yi Sung-ke had them all slaughtered, one by one. This gave rise to a gallows expression that still lives on in Korea: A "king's envoy to Hamhung" is someone who embarks on a trip and is never seen again.

While he was living in Seoul, Lee had sometimes wondered if that was what he'd become, in reverse—a sojourner who'd left Hamhung, not to be heard from again. A profound remorse had often tugged at him, a feeling that he was an impostor living a blessed life while his own family toiled and suffered in a homeland he would never revisit. He was convinced that he had abdicated his responsibilities, that he'd forsaken life's most important bonds.

Near the center of Hamhung, a major bridge spanned the Songchon River. Mansekyo, it was called, the Bridge of Long Life. At the Lunar New Year, people had made a tradition of walking across it—the exercise was said to bring longevity and good luck. Families would stroll back and forth, arm in arm, hand in hand, laughing and singing joyous songs. His own family had crossed the bridge each year—those outings were some of Lee's happiest memories. He wondered if there was anything to it. Had the Bridge of Long Life offered his family some balm of protection? Or was it only another naïve superstition?

When Japan took formal possession of Korea, in 1910, Hamhung was a medieval city steeped in just these sorts of myths and folk traditions. But in the mid-1920s, as the Japanese tightened their grip on the country, modernity began to arrive. A team of Japanese engineers struck upon an ambitious idea: They would build roads into the mountains northwest of Hamhung and harness the might of the Changjin River—Chosin in Japanese—an important tributary that flowed north toward the Yalu. In the highlands, some seventy road miles from Hamhung, the engineers would construct a large dam that would flood the valley floor. The Changjin waters would rise, swallowing the wrinkled country, and the resulting reservoir, with all its scallops and appendages, would extend southward for more than forty miles. It would be a deep lake splayed out in the mountains, practically on the rooftop of Korea.

This scheme alone was considered a nearly impossible feat, but then the engineers envisioned something bolder: They would effectively reverse the course of the river by building a network of pipes near where it entered the lake on its south end. The pipes would snake along, often underground, carrying cold lake water from the mountains to the coast. Thus, a river that had once flowed north would flow south, through man-made conduits. Working with gravity, these tubes of racing water would feed into a series of hydroelectric plants down on the plain that would supply Hamhung and its neighboring port city of Hungnam with enough power to transform the area into a military-industrial center, perhaps the largest in Korea.

Some said it was quixotic. Some said the engineers were tempting fate, manipulating sacrosanct forces of nature. But the immense project worked as planned. The Chosin Reservoir was completed in 1929, the year Lee was born, and, with dizzying speed, Hamhung-Hungnam underwent a metamorphosis, much of it under the direction of the Noguchi Corporation, a Japanese conglomerate founded by a chemical engineering mogul named Jun Noguchi, who was said to be the "entrepreneurial king of the peninsula." A nitrogen fertilizer plant, the largest in the Far East, was quickly constructed, and the area became one of the world's largest producers of ammonium sulfate. Then came oil refineries, chemical concerns, textile mills, metal foundries, munitions works. They produced dynamite and mercury oxide powder and high-octane aviation fuel. It was a grinding, stinking, spewing complex of industries designed to fuel Japan's expansionist aims across Asia.

Thousands of peasants, many of them displaced by the new lake, moved down from the mountains to work in the factories. Schools sprang up, a train station, a city hall, suburbs, all of it stitched together with streetcars and underground sewer systems and electricity and telegraph wire. It was a modern marvel of civil planning and central design—at least that was how the authorities portrayed the region's transformation. Through Japanese ingenuity and Korean sweat, men had built a lake that built a city.

≡

This was the boomtown atmosphere in which Lee Bae-suk had grown up. Throughout the 1930s, Hamhung quickly became, in many respects, a Japanese city—organized, industrialized, modernized, militarized. Korea was living under what came to be called "the black umbrella" of absolute Japanese rule. The occupiers humiliated and exploited Hamhung's citizens, often brutally, but they also sought to assimilate them—that is, to make them Japanese subjects, slowly eradicating all vestiges of Korean consciousness. As a boy in Hamhung, Lee was taught to bow toward the east, in the direction of the emperor. He prayed to Shinto gods, at Shinto shrines, kneeling in the shadow of red torii gates. At school, he and his classmates were required to recite the Pledge of the Imperial Subjects, promising to "serve the Emperor with united hearts." Lee, like all citizens, had to forsake his Korean name and adopt a Japanese one. He learned the Japanese language and was forbidden to study Korean in school. The Korean anthem was not to be sung, the Korean flag not to be unfurled, traditional white Korean clothing not to be worn. People were even expected to give up Korean hairstyles, cutting off their braids and topknots.

Everywhere Lee looked, he saw examples of Japanese authority and expertise: Japanese teachers, Japanese civil servants, Japanese soldiers and tax collectors and cops. The mayor was Japanese. So was the provincial governor. Even the city itself was given a Japanese name: Hamhung became Kanko. The Japanese Kempeitai, which many Koreans came to call the "thought police," tightened its hold on the city, stamping out dissent or expressions of Korean identity. The police organized the citizens into neighborhood associations, each one composed of ten families. These cells, designed to enforce compliance of Japanese laws, had a chilling effect on community relations, effectively turning Korean against Korean, requiring neighbors to spy on one another.

During the late 1930s, the industrial complex of greater Hamhung became an arsenal and a forge for Japan's deepening war against China. Enormous quantities of explosives were manufactured there. After Japan attacked Pearl Harbor, operations at Hamhung expanded exponentially. Among other secret projects, Japanese physicists made early attempts to build an atomic weapon. Using uranium reportedly mined from the mountains around the Chosin Reservoir, they constructed a crude cyclotron, produced heavy water, and even began to develop a primitive atomic device.

Lee and his classmates, though of course unaware of such covert doings, were inescapably pulled into the war effort. Throughout high school, they underwent compulsory military training, including sword and bayonet practice, in the expectation that they would soon fight against the Americans in defense of the Land of the Rising Sun. When he was fourteen, they put him to work in a refinery that made aviation fuel destined for use in the engines of Zeros. As the war progressed, the Japanese became more fiendish. Young Korean men were pressed into labor gangs and shipped to Japan to toil in mines, shipyards, and factories. Thousands of young Korean women were mobilized into something called the Volunteer Service Corps, ostensibly to work as nurses on the front. In reality, these women had been abducted to service Japanese soldiers in mobile brothels—"comfort women," these sex slaves were called. Rumors also floated that some Koreans were being used as guinea pigs in a sadistic medical experimentation program, based in Manchuria, known as Unit 731.

By August of 1945, Lee, then sixteen, knew that he would soon be conscripted to fight for the cause of a deeply evil regime. But after the Americans dropped atomic bombs over Hiroshima and Nagasaki, Japan surrendered. Lee was spared just in time. On August 22, the Soviet Army marched into Hamhung and liberated the city. Overnight, the once proud Japanese occupiers became refugees—chased from their homes, rounded up, and stripped of their assets. They became beggars, maids, ragpickers. Many were sent to Russian prison camps and never seen again.

At first Lee, like so many of his countrymen, was joyful that, thanks to world events, Korea had finally thrown off the yoke of her oppressor. But then a new reality began to settle in: The Soviets proved as ruthless as the Japanese, and the prospect of a free and independent Korea, that fervent hope that had animated generations of patriots, faded. Rampant were the stories of corruption, of executions without trial, of drunken Russian soldiers looting shops, ransacking homes, and raping Hamhung women. Traumatized girls and women began disguising themselves as boys and men in hopes of eluding sexual assault. The Russians would do anything for a drink, it seemed—they would down antifreeze until they vomited or break into people's homes in search of *sul*, a spirit distilled from rice. The Soviets systematically dismantled the factories of Hamhung and Hungnam, loading the most valuable equipment onto trains and hauling it off to Vladivostok. Such were the prerogatives of war, the spoils of a theater the Russians had entered only at the last opportune moment—spoils that they now expected to enjoy to the fullest.

Lee and his classmates went back to school. Only this time, instead of learning Japanese, they learned Russian. Stalin's face was everywhere, Cyrillic letters, banners colored blood red. Schools and buildings bearing Korean or Japanese names were instead given numbers. It became a crime to even speak the Japanese language, and baseball, a game that American missionaries and Japanese colonists had brought to Korea, was condemned as a decadent bourgeois sport. Talk of revolution hung in the air. Important Communists came from Pyongyang to deliver impenetrable, jargon-filled harangues, often in Russian, on the principles of Marxism and Leninism. Then followed the parades, the rallies, the self-criticism sessions. A new decree made it a crime for more than five people to meet without permission from the authorities. For Lee, the turning point came when he and a classmate attended a Communist assembly held in the public square, near the city hall and the train station. Posters of Kim Il Sung grinned at the crowds. On stage was a wall of implacable faces, a phalanx of new leaders in sharp military uniforms, smiling, waving. Everything had been decided—all posts filled.

Lee's friend, who was much more politically shrewd, couldn't contain his wrath. "Korea is not Russia!" he screamed. "Korea is Korea! One land, one language, one people!" He had yelled this message of dissent too loudly, for several military policemen had noticed him and were now coming for both him and Lee, pistols raised. The two boys looked at each other, wide-eyed, and took off through the crowd, running for safety.

This was Lee's initiation, the bud of his political awakening. He'd glimpsed the face of tyranny. Seeing no future here, he grew bitter and impatient. A line had been drawn at the peninsula's waist, and he yearned to cross over it. When he announced his intentions, his father, to Lee's surprise, assented. *You are the eldest: Go. Perhaps the others will follow.*

His father gave him some money and explained the arrangements: Lee was to take a train down to Wonsan, where a guide would lead him and a group of other young refugees south. They would lunch in temples and sleep in the tenebrous woods. They would eat food prepared by friendly co-conspirators. It was an underground railroad, following a network of trails that traced like capillaries over the hills and mountains.

One day in early 1946, at the age of sixteen, Lee embraced his mother and father. He gave his little sister Sun-ja a special hug. To his seven other siblings, he smiled an uncertain smile. Then he headed for the border, and a new life in Seoul.

Now, four years later, he was going home. As soon as X Corps had his papers ready, he would be flying to Hamhung.

11

HEROIC REMEDIES

Washington, D.C.

As the early-morning sun climbed behind the Capitol, President Harry Truman bounded out of Blair House, the executive guest quarters where he was living while the White House underwent renovations, and sauntered down Pennsylvania Avenue. He wore a natty suit and carried a cane. The day had broken unseasonably hot—temperatures were supposed to surpass eighty degrees by noon. Fall had become summer again. The birds appeared to stir with new vigor, and the last thumping insects of the season found they'd been granted a reprieve. Truman, too, seemed unusually sprightly on this day of Indian summer, November 1, 1950, as he bustled along the sidewalk.

The men who guarded the president hated these morning constitutionals. Each day he would veer in a new direction, winding through various precincts of Washington—Foggy Bottom, McPherson Square, the National Mall. He was famous for his brisk gait. The little man from Independence was a speed walker—120 steps a minute was said to be his pace. These strolls were a splendid way, said Truman, to get "your circulation up to where you can think clearly. That old pump has to keep squirting the juice into your brain, you know."

It had been two weeks since his meeting with MacArthur on Wake Island, and the trip's success threw off a rosy afterglow that continued to brighten Truman's mood. The president still had reason to believe that the Korean conflict was nearly over, that the troops, many of them, at least, would be home by year's end. The fond wish Truman had expressed at Wake—that Mao would officially stay out, that the conflict wouldn't escalate into a larger war—seemed to be coming true.

The president, having completed his walk, made his way to the West Wing, changed into his swimming trunks, and swam a few laps in the White House pool. He toweled off, did his usual fifty pulls on a squeaky rowing machine in the adjacent gym, and, after showering, dressed once again. At a little after nine o'clock, Truman entered the Oval Office and sat in the swiveling chair at his desk—decorated with a walnut sign that famously said THE BUCK STOPS HERE. Then he dove into his work.

After a staff meeting, Truman focused on a stack of papers that had been brought to his desk. One of the first documents that vied for his attention was a memorandum from General Walter Bedell Smith, the head of the Central Intelligence Agency. The memo concerned Korea, and it was a bombshell. It sprang not only from Douglas MacArthur's insular coterie of officers but also from the best multi-agency field reports, filtered through the best intelligence people in Washington. It was "clearly established," Smith's memo declared, that Chinese Communist troops had crossed the Yalu en masse and were flooding into North Korea. The Chinese had pushed as much as one hundred miles south of the river and were already "opposing" U.N. forces.

"Present field estimates," the CIA memorandum went on, were "between 15,000 and 20,000 Chinese Communist troops." This was an alarming number, but the CIA was not sure whether the primary objective of these Chinese soldiers was to attack U.N. forces or to defend a handful of hydroelectric plants that, though located in North Korea, generated power for southern Manchuria. Perhaps, wrote Smith, the Chinese sought to create a cordon sanitaire—a no-man's-land—south of the Yalu to protect these critical power plants.

Still, the CIA report spoke with the clarion voice of authority: "This pattern of events and reports indicates that Communist China has decided, regardless of the increased risk of general war, to provide increased support and assistance to North Korean forces." Smith suggested that the "Chinese Communists probably fear an invasion of Manchuria despite the clear-cut definition of U.N. objectives." The possibility cannot be excluded, he warned, "that the Chinese Communists, under Soviet direction, are committing themselves to full-scale intervention in Korea."

This was the first time the president had seen it definitively stated in writing: Mao's army was in Korea, clashing with Americans. The intelligence was real, and it was official—tens of thousands of Chinese Communist soldiers on North Korean soil.

=

Late that morning, Truman stepped outside the West Wing and made his way over to the Rose Garden in time to preside over a Congressional Medal of Honor ceremony. The president was bestowing the prestigious award on a Marine colonel named Justice Chambers, who had fought heroically at Iwo Jima. The day was gorgeous, and the lawn shimmered a radiant green against the white latticework.

The audience was a hive of military dignitaries, including Secretary of Defense George Marshall and General Clifton B. Cates, commandant of the U.S. Marine Corps. Whether the president took a moment to confer with either of the two men is not recorded. This was a celebration, not a time to brood or strategize, but the official news of Mao's entrance into Korea would have been a subject of much interest to both Marshall and Cates—the latter especially, since his First Marine Division was worming into the very frontier where increased Chinese activity had been reported.

Truman, who loved military ceremonies like this, realized that the atmospherics were a bit sullied by the major construction project that was making a ruckus just beyond the garden. The White House was little more than a Potemkin facade—gutted on the inside, its outer walls held up by an extensive network of girders. Earth-moving equipment clanked away, and workmen could be heard drilling and pounding deep within the mansion's cavernous hole.

When the president and his wife, Bess, had moved into the White House after FDR's death, they thought the place was haunted. Certain rooms, it seemed, were trying to talk. Floors heaved. Ceilings slumped. Crystal chandeliers tinkled overhead. In June of 1948, one leg of their daughter Margaret's piano broke through the rotted floor in her sitting room. Teams of engineers found crumbling masonry, sagging foundations, and fire hazards everywhere. One report noted that a number of beams "were staying up there from force of habit only."

So the Trumans moved out, and construction crews moved in. Deep in the bones of the building, contractors uncovered a bewildering palimpsest of old wiring, defunct plumbing, and ventilation ducts that hadn't been used in more than a century. The White House's "structural nerves" were seriously damaged, said one report. "Heroic remedies" would be required.

When it was learned, in 1949, that the Russians had successfully detonated an atomic weapon, the White House basement was dug deeper to accommodate a new hermetic bunker designed to withstand a nuclear blast. The modern world was changing at a frightening clip, and America, still new to her superpower status, struggled to retrofit herself for the perils of the role. For the time being, the most powerful man in the world's most powerful country, the man who had ushered in the atomic age, lived in what was little more than a glorified guesthouse across Pennsylvania Avenue.

After the Medal of Honor ceremony, Truman was driven over to Blair House for a simple lunch with Bess. Afterward, he went upstairs, stripped to his underwear, and opened the window. Then, as was his midday custom, he lay down for a nap.

≡

At 2:20 p.m., a White House policeman named Leslie Coffelt was pulling guard duty on the west side of Blair House when a twenty-five-year-old Puerto Rican man named Griselio Torresola snuck around the corner of the sentry booth, clutching a German Luger. Another Puerto Rican man, Oscar Collazo, armed with a Walther P38 semiautomatic pistol, approached Blair House from the other side of Pennsylvania. The two young conspirators wore chalk-striped suits and snap-brim hats. Neither of them bore President Truman any personal ill will—they hardly knew a thing about him, or his politics. But they were determined to murder him anyway.

Collazo and Torresola were Puerto Rican Nationalists, tied to cells that were attempting to foment a violent insurrection and assert independence for the island. The two men believed that only a sensational act would bring attention to their movement. They were also angry about the Korean War, and the contradictions they saw in the fact that so many Puerto Rican soldiers had joined the U.N. effort to fight for freedoms they themselves did not enjoy on their home island. If some of the two conspirators' grievances seemed reasonable, their plan was absurd. They had done no research. The plot had been hatched only a few days earlier. They'd taken the train down from New York the previous night.

Now Griselio Torresola wheeled on Officer Coffelt. In rapid succession, he shot the officer three times at point-blank range. Coffelt slumped in the chair of his booth, mortally wounded. Torresola then fired three shots at a plainclothes policeman, wounding him as well.

At another guard post, on the east side of Blair House, Oscar Collazo shot a policeman named Donald Birdzell in the leg, shattering his kneecap. Then a Secret Service officer opened fire, and bullets ricocheted off the pickets of Blair House's wrought-iron fence. On the busy sidewalk, bystanders scattered for cover, and a stray round smashed through the plate-glass window of a drugstore just down Pennsylvania.

President Truman, roused from his nap, dashed to the open window to learn what the commotion was. Still in his underwear, he stood squinting in the hot glare and looked down at the gunfight in progress. Had either of the conspirators looked up at the right moment, their ultimate target would have been standing in plain view.

"Get back! Get back!" someone yelled at the president, and he shrank into the shadows of his room.

Seconds later, Collazo was shot in the chest and fell, his body splayed at the base of the steps. He was critically but not fatally wounded. Then Officer Coffelt, bleeding in his sentry booth, summoned the energy to rise from his death throes. Holding his Colt revolver, he propped himself against the guardhouse and fired a single shot at Torresola, who fell dead behind a boxwood hedge, a bullet in his brain.

As abruptly as it had started, the battle stopped, and an eerie silence fell over Pennsylvania Avenue. It was the largest gunfight in the history of the Secret Service. Two men lay dead or dying, and three others were wounded. Twenty-seven shots had been fired in less than two minutes.

≡

An hour later, at the National Cemetery in Arlington, an assembly of dignitaries waited in the heat for the unveiling of a monument honoring a British field marshal who had been an influential adviser during World War II. Secretary of State Dean Acheson was sitting in the crowd when Treasury Secretary John Snyder hurriedly took a seat beside him. "An attempt has been made to assassinate the President," said Snyder, alarm in his voice. "Shooting has been reported in front of Blair House."

Acheson asked if Truman was okay.

"I don't know," Snyder replied.

Others around them had overheard the exchange, and now murmurs and gasps worked their way back through the crowd. Then the Marine Band struck up "Hail to the Chief." The president's limousine pulled up, with eight Secret Service agents following behind in a large open car. Truman stepped out, looking stern but calm in a dark blue suit. An extra detail of security officers was posted throughout the crowd, and other agents were seen lurking behind trees in the cemetery.

By the time Truman reached the assembly and took his seat, he wore a grin and looked, thought Acheson, as if "he had not a care in the world." The secretary breathed a sigh of relief. "All of us in Washington," he said, had suffered "a bad scare."

Later, when asked about the attempt on his life, Truman was pragmatic. "A president has to expect these things," he said. "But it was a most unnecessary happening." His thoughts were with the "grand guards" who had saved his life—and especially with Officer Coffelt, who had died in the hospital.

Truman could not hide his contempt for the would-be assassins. "The two men who did the job were just as stupid as they could be," he said. "I know I could organize a better program than the one they put on. One of them faces the gallows, and the other one is dead."

After the ceremony, Truman spoke briefly with George Marshall, then slipped into his limo and was whisked across the Potomac. By early evening, when Truman returned to the Blair House steps, a cleaning crew had scrubbed away the bloodstains.

12

WILL O' THE WISP

Sudong Gorge

The village of Sudong was a tiny collection of mud-and-wattle huts on the main road leading up to the Chosin Reservoir. It was the gateway to the Taebaek Mountains, the place where the coastal plains petered out and the highlands began. The surrounding countryside was a patchwork of now brittle rice paddies and spindly persimmon trees. This late in the season, most everything had been harvested, and the farmers were hard at work preparing their foodstuffs for the winter. Oxcarts trundled along the berms that divided the fields. A crisp exigency informed the villagers' movements. They seemed to be preparing for something—for the coming cold, and for the coming hostilities.

By November 2, most of the three thousand Marines of the Seventh Regiment had moved up to Sudong. The troops, having bivouacked for several days in a rat-infested warehouse in Hamhung, were pleased to make this bucolic place their new home. To the south of the village was a deep gorge crowded with boulders, and it was here, beside a dry riverbed, that much of the regiment pitched camp. Most of the men seemed carefree—insouciant, even. They believed the scuttlebutt they'd been hearing: that they were headed home soon, that it was all sewn up, that this was just a final, rugged exploit in the country.

By now everyone had heard that the Chinese were supposed to be gathering in the hinterlands hereabouts. But so what if they were? After those two wretched weeks trapped aboard ship, many of the Marines were spoiling for a fight, spoiling for action of any kind. The prevailing attitude seemed to be: If the North Koreans had vanished into the hills, then bring on the Chinese. How disconcerting it would be to come all this way and never see an enemy.

The day was bright and full of promise, and the upland air smelled of pine. Perhaps the Marines had been lulled by the uneventfulness of their entry into North Korea. They had trucked and marched up from the coast without encountering anyone—save for the kids who lined the road, flashing wide smiles and fluttering little flags. Wrote one Marine: "Korea began to look like a phony war. A spirit of confidence, a false sense of easy victory, began to take hold."

Lieutenant Joe Owen recalled the chipper mood on the day they made for Sudong. "There was great energy in the ranks," Owen wrote—they were "healthy young men on the way to adventure." Owen's unit, Baker-One-Seven, happily bedded down in an apple orchard. He and his mates cleaned their weapons and whetted their bayonets, and the "old salts retold stories of Japanese ferocity on the Pacific islands." The men sang "ribald songs" and went out to a meadow to play a game of football—in lieu of a pigskin, they used a squashed-up jacket cinched tight with a web belt. It felt almost like a hunting trip—young men camped in the boonies, smoking and bullshitting around fires, fiddling with motors and guns. A kind of Indian summer had come to North Korea, with unseasonably warm temperatures, and this put the men in a frisky mood—some even skinny-dipped in a nearby stream.

But the commander of the Seventh Regiment, Colonel Homer Litzenberg, scowled at this lightheartedness. He had a sense that something momentous was about to happen. Litzenberg, who went by the nickname Blitzin' Litzen, was a gruff, no-nonsense, square-jawed Pennsylvanian of Dutch descent who, during World War II, had served at the battles of Tinian and Saipan. One officer who served with him in Korea called him a "very stubborn Dutchman" who was a "bit of a bully," but it was also said that he could be moved to tears by stories of the suffering of his men. He was a "tall, ruddy-faced man, built all of rectangles and squares." Like General Smith, Litzenberg had prematurely white hair, which gave rise to his other nickname among the Marines: the Great White Father.

Litzenberg took his regimental officers and noncoms aside and gave them a sobering talk. One day soon, he said, they might be fighting the first engagement of World War III. "We can expect to meet Chinese Communist troops," he warned, "and it's important that we win the first battle." He told his officers to pass the message down through the ranks: Facing the Red Chinese would be nothing like facing the threadbare and skittish North Koreans. It wasn't just that Mao's troops would likely be better trained, better organized, more experienced; it was also the weight of the culture behind them, the might of an ancient society, the pressure of a virtually numberless people. Litzenberg was convinced that his regiment would see the opening salvo in the next global conflict, a collision of ideologies that would be felt across years and continents. "We want the outcome to have an adverse effect on Moscow as well as Peking," Litzenberg said. "The results will reverberate around the world."

The Great White Father's words had the chilling effect he intended, and the essence of what he'd said filtered among his battalions. That night, under a starless sky, the men shimmied into their bags and fell uneasily to sleep.

≡

And then, just like that, it began. Close to midnight, the men heard a cacophony of bugles and horns, and the Red Chinese fell upon the gorge—"flights of them," said one account, "like flocks of blackbirds." They attacked the Seventh Regiment's two leading battalions and infiltrated the gap between them. A battle raged through the night and, in fits and starts, for the next several days. By the time it was over, sixty-one Marines, and an estimated one thousand CCF soldiers, had been killed. Smith, often prone to understatement, called it "quite a fight."

But almost as quickly as they had appeared, the Chinese vanished. They withdrew into the mountains to the north, and the Sudong Gorge fell silent. Colonel Litzenberg and General Smith were left to wonder why. What strategy were the Chinese employing? They appeared to be seeking intelligence of some kind, not terrain.

And judging by the erratic way they had fought, and by their sudden withdrawal, they seemed less interested in achieving a battlefield victory and more interested in sending the Marines a message. But if so, what was it?

Was this attack merely a shot over the bow to issue a warning? *We are here! Come no further!* Was it an attempt to probe the Marines, to assess their strength and test their will to continue? Was it a blocking and delaying action, sacrificial in nature, to buy time for larger and more powerful Chinese units to arrive from Manchuria? Or was their appearance at Sudong a face-saving demonstration to fulfill a political promise the Chinese may have made to the North Koreans—a token effort to at least appear to help their Communist brothers to the south?

Another possibility occurred to General Smith: Did the CCF presence at Sudong have something to do specifically with the Chosin Reservoir itself? Perhaps they'd been sent down to guard the mountain gate that led to this important hydroelectric concern, which supplied power to grids that fed deep into Manchuria. This brought Smith back to a point he had puzzled over since the first time he heard about Almond's plan to advance to the Chosin Reservoir and other hydroelectric installations to the north: Why, of all the places on the map, were they driving straight for the cluster of assets that the Chinese had indicated were of vital strategic importance to their economy? This seemed to Smith like a deliberatively provocative move on the part of the United States—or at least one that the Chinese could not help but perceive in that way.

Then again, both Smith and Litzenberg wondered, maybe the Chinese generals were trying to accomplish something far more cunning at Sudong. The attack could have been a stratagem designed to make the Marines think they were weak, to coax them into advancing further and deeper into the mountain wilds, where the Chinese might more easily envelop them. Had Litzenberg been right from the beginning? Maybe this *was* the start of World War III, and the worst was yet to come.

It was possible that some combination of all these explanations was at play. The enemy's true motives and objectives at Sudong remained a mystery that would be debated for decades. And yet the Chinese prisoners captured at Sudong were just as forthcoming as the ones the South Korean troops had taken a few days earlier. The Red Chinese readily told Marine interpreters that they were from the 124th Division, and that they had crossed the Yalu in mid-October. They wore quilted off-white uniforms, and seemed docile and eager to please. They hinted that many hundreds of thousands of troops were behind them: The 124th was only the first wave. The interpreters surmised that this could well have been a script the Chinese troops had been instructed to recite if captured—a tale to spook the Americans and make them think twice about continuing. On the other hand, the prisoners could have simply been telling the truth.

One of the Chinese prisoners caught near Sudong made a lasting impression on the Marines. He was "a tiny fellow who smiled continuously," recalled one Marine battalion commander. "He was very hungry and wolfed down the C-rations we heated for him." The Marines then offered the captive a sleeping bag, and he fell sound asleep in the middle of the battalion command post. When he awakened, the prisoner asked through an interpreter if he could return to the mountains and gather some of his company mates—who, he insisted, were most eager to surrender. A Marine officer assented to the request and lavished the young Chinese man with supplies of food, cigarettes, and other enticements. Escorted to the Marine perimeter, he seemed in a jolly mood as he scampered up the ridge. At the top, he turned and gave everyone a wave. Then he vanished into the fog. He was never seen again.

General Smith, from his Hamhung headquarters, pondered the battlefield situation anew. He had no illusions about the foe he was facing—"we are fighting a sizeable unit of the Chinese Communist Army," he wrote. But where were they now? The spotter planes were able to glean little solid information from the air. It seemed the 124th had evaporated into the mountains. ("It was the quiet that worried them," said one Marine account. "An army as big as the Chinese ought to make noise.") Smith took a helicopter to Sudong to confer with Litzenberg. The two men found it cryptic. "It was very difficult to obtain an accurate picture of the enemy situation," Smith wrote. The Chinese, he said, were "like a will o' the wisp."

≡

Whatever had happened at Sudong, whatever larger lessons might have been learned, Ned Almond didn't consider the engagement worthy of contemplation or analysis. The X Corps general didn't mention the battle in the command diary he regularly kept. Still, Almond went up to Sudong and, with an interpreter, interrogated some of the Chinese prisoners himself. To him, they seemed pitifully trained, poorly equipped, and scared—a band of amateurs. They weren't anything to worry about. He described the enemy as merely a "bother." However, there was no doubt in Almond's mind that these were, in fact, Chinese, and he radioed Tokyo to alert MacArthur of their presence. MacArthur, in turn, dispatched his intelligence chief, General Charles Willoughby.

Willoughby was a strangely formal native of Prussia. Born Adolf Karl Tscheppe-Weidenbach, he was a secretive man, and much about his German past was shrouded in mystery. MacArthur affectionately called him "my pet fascist." (An extreme right-winger, he later became an adviser to Spanish dictator Francisco Franco.) General Willoughby ran his intelligence shop in a spirit of splendid intrigue, and always with the goal of pleasing the supreme commander. "Anything MacArthur wanted," said one Army colonel, "Willoughby produced intelligence for." Subordinates described Willoughby as a sickly, misanthropic loner who spent inordinate amounts of time at the movies and devoted most of his energies to currying favor with MacArthur. "He has no wife, no family, and lives for himself alone," wrote Lieutenant Colonel James Polk, an intelligence officer who worked directly beneath Willoughby in Tokyo. "I don't think he has a real friend in the world."

Willoughby quickly concluded that the Chinese prisoners at Sudong were merely "volunteers," part of a token force of zealous Communists, probably from Manchuria, who had picked up their weapons and, in piecemeal fashion, streamed down of their own free will to help the North Koreans. This, in fact, was the story that Peking Radio had been broadcasting and that representatives of the Chinese government, through various channels, had announced to the world.

It was a fiction that MacArthur seemed entirely willing to believe, one that provided sufficient cover for him to continue advancing toward the Yalu. The Joint Chiefs of Staff had given him carte blanche to keep on going—unless and until he saw evidence that the Chinese had officially entered the war. And here it was, evidence as clear as one could ever expect to find. MacArthur's response was to accept Beijing's propaganda at face value.

Tokyo's failure to fully recognize the Chinese presence was all the more extraordinary given the critical events that had simultaneously occurred on the west side of the Korean Peninsula: While the Marines were engaged at Sudong, another massive Chinese force had attacked South Korean units, as well as the U.S. Army's Eighth Cavalry Regiment, at Unsan, in northwestern North Korea. United Nations forces suffered more than a thousand casualties at Unsan. Here was further proof that the Chinese were truly committed—proof that Charles Willoughby preferred to ignore.

A number of intelligence people in Tokyo would later allege that Willoughby suppressed field reports to conform to the strategic picture MacArthur preferred to believe. "We had the dope," insisted Lieutenant Colonel James Polk. "The fault lies in that Willoughby didn't insist on the truth of his dope. He bowed to the superior wisdom of his beloved boss and didn't fight him as a good staff officer should." Instead, thought Polk, Willoughby was more concerned with feeding MacArthur's ego and "vying for the favor of the most high."

How Tokyo had arrived at its conclusions, General Smith could not understand. These were no "volunteers." The Chinese had fought strangely at Sudong, but they had fought well. They were regular army troops, all right—Mao's troops. Many of them reported that they hailed from places deep in the hinterlands of China and that they had been lifelong professional soldiers. They had journeyed far, and at someone's considerable expense, to get here. They had not come out of the kindness of their hearts to expel the imperialist invaders.

The Chinese at Sudong had sent an important signal, but the American high commanders had missed it. Alexander Haig wrote that Almond found the Chinese behavior "puzzling . . . the enemy was too subtle for us, and we were too obdurate for them." By evaporating into the ether the way the Chinese did, they enabled Almond to entertain a false sense of victory, fueling the illusion that his X Corps had caused the enemy to scatter. In this way, Sudong led to yet another miscalculation in a host of miscalculations.

But with this difference: Americans had spilled blood and died on a battlefield at the hands of the Chinese. The stakes were no longer abstract; United States troops were committed now, not just strategically but emotionally. Much was made of the "Oriental" need to save face, but Americans were not immune to the phenomenon. They had to protect their investment, to guard the national prestige, to honor the men who had fallen.

On November 9, Almond ordered Smith to start moving his Seventh Regiment from Sudong farther up the mountain, to a little place called Koto-ri. Almond's ambition to reach the Yalu River remained undiminished, for now he wanted Smith to make for the Chosin Reservoir and assess the situation from there.

Smith passed down the orders reluctantly, and in so many words told Litzenberg to ignore Almond's call for haste. Contrary to his name, Blitzin' Litzen was to proceed slowly and with maximum caution. Smith's operations chief, Bowser, said Smith wanted the Seventh Marines to pull "every trick in the book to slow down our advance, hoping the enemy would show his hand before we got even more widely dispersed than we already were."

On the morning the Marines struck their camps and wound their way up through the pass and onto the plateau, most of them felt a tinge of apprehension, a dreadful tingle in the belly. The known presence of an officially nonexistent foe led some to utter a kind of tautological jest: *If you are shot and killed by an enemy that is not there, are you still alive?* Some tried to relieve the jitters with gallows humor and assorted vulgarities. "We joked and laughed as we marched," said Lieutenant Joe Owen, "and made obscene comments about the things that were central to our lives: the chow, the terrain, the enemy, the lack of women." *When I get home to my wife,* they'd say, *the second thing I'm gonna do is take my pack off.*

As they rose into the mountains, the landscape turned as gray and somber as a Rembrandt painting. "Even at noon there was shadow," said one Marine account. "Shadow and shade, a gloom, a darkness, over the snow and the land." You could scarcely reach a summit, it seemed—the forbidding terrain rolled ever upward. Said Colonel Litzenberg: "Beyond each hill lay another hill, and that one always seemed higher." This kind of real estate, said General Smith, "was never intended for military operations. Even Genghis Khan wouldn't tackle it."

The narrow dirt road was badly rutted and did not seem to have a name, or even a number. The Marines called it the MSR—the main supply route. It corkscrewed as it climbed into the foothills, and in places it had been carved implausibly into the side of the mountain, in seeming rebuke of the forces of gravity. With every forward step into this stark domain, the men of the regiment felt they were getting more deeply enmeshed in a scenario they could not escape. Meat on a stick, they called themselves. Easy pickins. As they drove deeper into this "bandit country," wrote one Marine, "the risk of ambush grated and rasped and the men turned edgy." The road, said another, seemed to be leading them into a "mysterious Oriental kingdom. I expected to see a giant ogre lurking on the sawtooth horizon."

13

BROKEN ARROWS

Washington, D.C.

I f the meaning of Sudong was lost on Almond and MacArthur, so also was it lost on President Truman and his advisers in Washington. Over the past four days, Truman's attention had been diverted to other arenas. It had been an extraordinary passage, packed with consequential developments. On November 7, midterm elections were held across the nation, and Truman's Democrats were deeply disappointed by the results. The Republicans gained twenty-eight seats in the House of Representatives and added five new seats in the U.S. Senate. In many campaigns around the country, the Korean War was a decisive issue—the Truman administration came in for widespread criticism for its prosecution of a conflict that was controversial and unpopular.

Also looming over the election were the charges leveled by an obnoxious young senator from Wisconsin named Joseph McCarthy, who seemed to find the color red beneath every stone. Back in February, McCarthy had risen to national notoriety with a sensational accusation that Secretary of State Dean Acheson had knowingly permitted his State Department to become "infested" with more than two hundred card-carrying Communists. McCarthy warned that certain "enemies within" had undermined the nation's security. Calling for loyalty tests and other extreme measures, he accused Truman of being in league with known Communists and charged that the Democratic-held White House had presided over "twenty years of treason." McCarthy would say of Truman: "The son of a bitch should be impeached."

During the weeks leading up to the 1950 election, McCarthy had campaigned on behalf of Republican candidates all over the country. Although he was a divisive and often outrageous character, his Red-baiting fulminations often fell on receptive ears. America in the fall of 1950 was immersed in Cold War paranoia, swept up in nuclear hysteria, gripped by Russian fever. The Iron Curtain had descended, and it seemed as though every month yet another Eastern European country had buckled to Stalin's Soviet Union—a Soviet Union that now had atomic weapons. In the binary world of Communism versus capitalism, Communism seemed to be winning. The environment was ripe for any demagogue sly and brazen enough to exploit it. Predictably, McCarthy inveighed against the Truman administration's creation of a "Korean death trap," saying that "we can lay [it] at the doors of the Kremlin and those who sabotaged rearming, including Acheson and the President."

Already, 1950 had produced enough political drama to leave the perception of a cumulative taint on Truman's tenure. It was a time of fears and doubts; the country seemed to be jumping at its own shadow. American news was full of espionage scandals and charges of conspiracy. Earlier that year, Alger Hiss, an American diplomat who had been accused of being a Soviet spy, was convicted of perjury in a widely publicized trial. In England, a German-born physicist named Klaus Fuchs, who had worked at the Los Alamos Laboratory during World War II, confessed to having supplied secret nuclear information to the Soviets. Then there was the controversial case of Julius and Ethel Rosenberg, the Jewish American couple who had been indicted for treason back in August and were awaiting espionage trials that would lead to their executions.

None of these developments had anything in particular to do with Truman, of course, but a pall of allegation and intrigue seemed to hover over his administration. The perception of dominoes falling, of containment policies failing to contain, of Reds permeating the halls of government, gave GOP contenders much ammunition with which to mount their 1950 campaigns against the Democrats, who were said to be soft on Communism.

It deeply worried and saddened Truman that McCarthyism

seemed to have had a national electoral effect. On some level, the senator's reckless tactics and innuendos had *worked*. But if anything, the demoralizing results of the midterm election only made Truman double down on his commitment to winning in Korea—and to proving his critics wrong.

≡

Two days after the election, on November 9, an incident over the Yalu River captured the full attention of the Truman White House, and the Pentagon as well. A U.S. Navy fighter pilot, flying a Grumman Panther, was on a mission to bomb several bridges near the river's mouth, at a place called Sinuiju. The pilot of the jet, Lieutenant Commander Bill Amen, had taken off from the aircraft carrier USS *Philippine Sea* and was flying over the river when he encountered an enemy aircraft boring in on him: a Soviet MiG-15 that was piloted, it would later be learned, by a Russian captain, Mikhail Grachev.

That argentine flash in the sky must have been a forbidding sight. At the time, American experts believed that the swept-wing, snub-nosed MiG outclassed any fighter jet the U.S. armed forces had in the air. But no one knew for sure, for no American jet had ever fought a MiG before. Now, over the Yalu, a genuine dogfight commenced, with the two sleek warplanes darting through bands of low, hazy clouds. Lieutenant Commander Amen, circling around and closing in on Grachev, scored a mortal wing hit on the MiG, which went into an inverted dive and crashed to earth. Amen safely returned his Panther to its carrier.

The incident was not widely reported, but it made history: It was aviation's first jet-on-jet kill. It also provided hints that the vaunted MiG was not invincible. But in Washington and in Tokyo, this isolated victory provided little cause for celebration. If an American pilot had won the opening round in a new hyper-adrenalized kind of aerial warfare, that was fine, but American commanders considered it a portentous day nonetheless. The Soviets had stationed a number of MiG squadrons at a Chinese air base in Antung Province, Manchuria, and now it was clear that they intended to use them. In the days ahead, encounters with Soviet jets would become so numerous that this stretch of the Yalu would acquire a new nickname: MiG Alley.

While Stalin insisted that only North Korean pilots were flying the MiGs, American fighter pilots suspected otherwise. Among other telltale signs, they swore they heard Russian voices crackling over the radio waves.

If this were true, then President Truman had larger implications to mull over. Not only had the Chinese actively entered the war; now, quite possibly, so had the Russians.

≡

The following day, November 10, an even weightier event took place that would again disturb Truman's concentration. That night, a Boeing B-50 Superfortress took off from Goose Bay Air Base, in Labrador, Canada. Flying over the St. Lawrence River, the heavy bomber ran into trouble. First one and then another of its four engines failed. Protocol required that the pilot immediately jettison his cargo—and so he did, right over the river, not far from the city of Rivière-du-Loup, Quebec, 250 miles northeast of Montreal.

The cargo in question happened to be a Mark IV atomic bomb, a revised version of the "Fat Boy" that had obliterated Nagasaki five years earlier. The crew set the squat, five-and-a-half-ton device to detonate at an altitude of 2,500 feet. Mercifully, the bomb was missing its plutonium core, so no nuclear reaction occurred. But the resulting explosion was massive nevertheless, and it rained more than a hundred pounds of moderately radioactive uranium over a wide arc of the Quebec countryside. The shuddering blast woke residents on both shores of the river for many miles. Soon afterward, the stricken bomber managed to land at Loring Air Force Base, in Maine.

American and Canadian officials immediately moved to cover up the accident, telling reporters that what residents had heard was merely a five-hundred-pound "practice" bomb—conventional, not atomic—that had been intentionally and safely detonated. Not until the 1980s would the United States Air Force acknowledge that this was a case of a lost nuclear bomb—there would be several during the Cold War—an incident category known in military parlance as a "broken arrow."

For President Truman, it was yet another distraction, another twist, in a spectacularly nerve-racking week.

14

A POWERFUL INSTRUMENT

Hamhung

On the night of November 10, in the mess hall of the division's new Hamhung headquarters, a group of officers gathered around General Smith as he grasped an old samurai's sword. With an air of precision and ceremony, Smith raised the blade in the air, then brought it down upon a birthday cake. The baked concoction was nothing special, a slightly misshapen affair the Marine cooks had spontaneously whipped up, slathered in chocolate icing. A stack of plates and forks was set nearby, along with a triad of votive candles and a pitcher of punch.

The birthday this solemn party of men had come to celebrate was not Smith's—nor was it the birthday of any officer now crowding around the table. Rather, it was the anniversary of the founding of the United States Marine Corps. In loopy lettering the bakers had squeezed from a piping bag, the top of the cake read U.S.M.C., 1775–1950. On this day, the Marines had turned 175 years old.

The Corps was diligent about commemorating its natal day. The Marines may have had a reputation for being crabby warriors with hearts of iron, but they could be sentimental. There was an old tradition in the Marines, wherever they might be, whether at home or on the most trying battlefield, to commemorate the Marine birthday with a cake and a corny little ceremony like this one. As the story goes, the Marine Corps was established on November 10, 1775, in a tavern in Philadelphia—Tun Tavern, it was called. The details are sketchy, but that was when and where the first recruits of the Continental Marines were said to have formally enlisted during the Revolutionary War.

This was the hallowed occurrence General Smith was determined to observe—and not just here in Hamhung. Smith had made sure that his regiments, scattered in their encampments out in the field, had taken a moment to recognize the day with a few solemnities and a freshly baked cake—or at least the best approximation the cooks could improvise from their mess tents.

Smith cut the cake with the rusty sword he'd been handed and distributed the pieces according to custom: the first slice to the oldest Marine in the room, the second slice to the youngest. He recited passages from the Marine Corps Manual and then read greetings that had been sent in from various dignitaries, including this one, from Vice Admiral C. Turner Joy, commander of naval forces in the Far East: "On the occasion of your 175th Anniversary I consider it indeed an honor to salute your courageous comrades in arms. You can justly be proud of your past record, and of your present gallant and heroic exploits in the Korean Campaign."

Then the cake and glasses of punch were served. The ceremony was finished in a half hour, and Smith and his staff scattered to their bunks.

$$\equiv$$

One hundred seventy-five years of the United States Marines: It was a trivial thing, perhaps. But celebrating the day was one of the rituals by which Marines reminded themselves of their peculiar place in the world and their sometimes tenuous perch within the U.S. military. The Corps had long possessed a strain of elitism. The Marines believed they could do more with less than any other fighting force on earth. At times they seemed almost like a cult. "We're a ferocious little confraternity . . . a violent priesthood," wrote James Brady, a Korean War Marine. "You aren't simply enrolled, but ordained." The Marines were steeped in lore, having led such legendary campaigns as the Battle of Belleau Wood, in World War I, and Iwo Jima, in World War II. They were bullish about their props and mottos, their traditions and regalia—the bulldog mascot, the anchor-and-globe insignia, the mawkish anthem, which they loved to sing at every opportunity.

The Marines had their own structures, their own slang. They were jarheads, leathernecks, devil dogs. Just don't call them soldiers. They made an odd grunting noise—*Oorah!*—which was their battle cry. They were proud and boastful, and disparaging of rearguard Army troops, whom they called "doggies" and other derogatory terms. From the halls of Montezuma to the shores of Tripoli, the Marines believed they were imbued with special traits, a special mystique. They were the first to fight, and they were always faithful: *Semper fi.*

Outsiders could find their bragging insufferable, but Marines had good reason to think and talk this way. Company for company, platoon for platoon, the Marines had established a long track record as the most effective—and the most lethal—troops in the United States armed forces. It was hard to say why this was so, hard to identify the exact quality or set of qualities that made the Marines so efficient in the heat of combat.

Martin Russ, a Korean War Marine who later became an accomplished writer and historian, put it this way: "It was not because they were braver or had God on their side; it was because Marine recruits were inspired from the beginning with the conviction that they belonged to a select and elite legion, and because of a tradition of loyalty which meant in practical terms that the individual trusted in and relied on his comrades to an extraordinary degree, and that he himself was trustworthy and reliable. Most Marines of that day believed it was better to die than to let one's comrades down in combat. The ultimate payoff of this esprit de corps was a headlong aggressiveness that won battles."

It was Marines serving in China during World War II who took a local industrial expression and popularized it, creating what became the definitive adjective capturing the special quality Marines were supposed to have: *gung-ho.* Its Chinese characters literally meant "work together."

Perhaps in part because the Marines had no elite academy to define and foster their officer class, the ethos of the Corps tended to be egalitarian, from top to bottom. This was best reflected in their most fundamental mantra: "Every Marine a rifleman." The expression meant that all Marines, no matter their rank or specialty or job description, were supposed to know how to wield a weapon and fight with the lowliest grunts. Marine bakers and Marine radio operators were first and foremost riflemen—and so were Marine generals. An "I am Spartacus" camaraderie was at work here; when the fighting was fiercest, every Marine was supposed to be interchangeable, and equal before the exigencies of battle.

The Marine style was no-frills, nothing showy. Their uniform was a basic forest green, without embellishment. This was a point of confusion for some people who saw them only in formal garb, on ceremonial occasions: Marine Corps bands strutting to John Philip Sousa, or those immaculate creatures who guarded embassies—white-gloved mannequins with high-and-tight buzz cuts, their dress blues edged in blood stripes. But most Marines said they preferred to be at the front lines, in foxholes, fighting "our country's battles," as the hymn put it, and keeping "our honor clean."

For all their esprit de corps, the Marines also suffered from a persecution complex, a sense of wounded pride, a feeling that the military-political hierarchy in Washington misunderstood and underappreciated them. At times the Marines wondered if they were the redheaded stepchildren of the military. They were neither fin, nor fur, nor feather. Though formally attached to the Navy, they weren't sailors. One could think of them as infantry, but they were emphatically not of the Army. They had their own cadre of excellent aviators, but they were not Air Force. Nor were the Marines considered "special operations forces," like the Rangers or the Green Berets, or any other units schooled in stealth and the raiding arts.

The point was, the Marines were their own tribe, a breed apart. This made them outcasts of a sort, a status they begrudged but also savored. They thrived on the very thing they resented. They felt they were constantly having to prove their worth, that in the public's eye they were only as good as their latest exploit. On the battlefield, they tended to adopt an orphan's mindset: *No one can save us but ourselves.*

This chronic grievance had recently been reinforced by none other than President Harry Truman, no fan of the Marines. In a letter leaked to the press only a few months earlier, Truman had belittled the Corps as "the Navy's police force," adding that "as long as I am President that is what it will remain." The Marines, Truman went on to say, "have a propaganda machine that is almost equal to Stalin's." Truman had roundly apologized, but his remarks were taken as a deep insult by many Marines, including Smith. Before the outbreak of the Korean War, it had been bruited about that the president (an Army man all the way) wanted to decimate the Marines as a fighting force—or disband them altogether. Many Marines in Korea had the sense that they were not only fighting the enemy, but fighting to save themselves as an institution. The perception was that their very existence was on the line.

=

That night, after the cake was eaten and Smith and his staff had gone to bed, winter arrived with a vengeance. There was an old saying, apropos of cutting cakes with swords, that the calendar year in North Korea had no swing seasons: When summer ended, winter fell "like a samurai's blade."

Perhaps the Indian summer of the past week had made the men apathetic, had caused them to forget the warnings they'd all heard. Of course, they'd been told that Old Man Winter was no joke in this part of the peninsula. The weather was supposed to be like that of upper North Dakota, or Saskatchewan. But the men weren't prepared for such an enveloping chill, neither its intensity nor the abruptness of its descent. Overnight, temperatures went into free fall. Within a few hours, the mercury plummeted forty degrees, to nearly ten below zero Fahrenheit. Then came the winds—flaying, buffeting, straight from the steppes of Manchuria. The gusts built in waves until they measured thirty knots.

The effect on Smith's men was devastating, particularly for those in the Seventh Regiment, who were now garrisoned high in the mountains, in the village of Koto-ri. Litzenberg's camp was as bleak as a Siberian ice station. One Marine journalist described the cold there as "wet, raw, devouring . . . a howling beast." The chill was powerful enough, said an official history, "to numb the spirit as well as the flesh." It stole into men's nostrils, took away their breath, froze the phlegm in their sinuses. Spitting resulted only in "a disappointing crackle" upon the ground. "The cold was a physical force you had to reckon with," said a Marine MP from Massachusetts. "It got down into the marrow of our bones."

The chill altered people's personalities, too—it made chipper guys brood and tough guys cringe. It was so frigid, said one account, that braggarts could "find no breath to boast in." It made people daft. They stopped talking, said Lieutenant Joe Owen, "except to blaspheme the goddamn fools who sent us out into this miserable, cold country." "If I'd known what the temperature was," said a truck driver from North Carolina, "I probably would have died."

At the mess tent, a cup of scalding coffee would accumulate a skim of ice within minutes. Canteens and C rations froze solid. Fingers stuck to metal. Helicopters refused to rise. Truck engines balked. Rifles seized. Batteries fizzled. The cold seemed to come with only one upside: It had a cauterizing effect on wounds. Blood from bullet holes or shrapnel tears simply froze to the skin and stopped flowing.

The quartermasters had already issued the men warm winter clothing—windproof trousers, alpaca-lined parkas, heavy woolens, mountain sleeping bags—but the gear was inadequate to combat this kind of cold. In Koto-ri, warming tents were set up, with kerosene heaters roaring night and day. Some men, however, needed more than simple warmth; they needed to go straight to the infirmary. Dozens of Marines had collapsed as though from exhaustion. They seemed to be in shock. They were dazed, semiconscious. Their vital signs became erratic, and their respiratory rates dropped to dangerously low levels. Others suddenly transitioned from this catatonic state into a hysterical sadness, sobbing uncontrollably. The Navy medical corpsmen developed a term for these extreme cold casualties: They were "*shook*," they said, a description that seemed to get at both the physical and the psychological aspects of their condition. General Smith noted in his log that "stimulants were required in addition to warming in order to restore the men to normal."

"The Marines' Hymn" boasted that the leathernecks had fought "in ev'ry clime and place where we could take a gun," including in "the snow of far-off Northern lands." Wintry North Korea qualified as such a place, but in his substantial reading, Smith could not recall the Marines having fought a battle in conditions quite like this. General Smith, a Californian in every sense, had spent his life in sunny places whenever he could help it. He never mentioned it to anyone, but he had a long-standing physical aversion to cold weather: When the temperature dropped, he experienced tremors in his hands and a numbness in his feet. The symptoms, he thought, dated back to a nearly fatal case of influenza he contracted in 1918. Those symptoms, in turn, had become exacerbated during a long winter he served in Iceland early in World War II. Ever since then, he'd tried his best to avoid getting chilled. Peleliu and Okinawa, where he later fought, were both horror shows in the extreme—but at least they were warm.

Even worse, this was the beginning of a peculiarly severe weather pattern that would lead to one of the coldest North Korean winters on record. More than Smith realized, the weather would not only be a factor in the coming conflict; it would be, in effect, the third combatant, an omnipresent force that would color his every decision and follow his every move. The farther his men traveled from the coast and the higher they climbed into the mountains, the stronger this third combatant would become. "It seemed with each step," said one account, "the air was changing, cooling, closing its grasp on the earth." Few American armies—few armies from any nation—had ever conducted extended operations in such harsh, sub-zero conditions—in an alpine landscape, no less. Wrote Martin Russ: "General Winter, having won many a campaign down through the centuries, was about to reap a heavier harvest of casualties at the reservoir than the armies themselves with their bombs and bullets."

=

General Almond, working mostly in the warmth of his X Corps headquarters in Hamhung, was unfazed by the arctic weather the troops were experiencing in the mountains. On November 11, the day the temperatures dropped so precipitously, he passed down an extraordinary order. The old plan, which had momentarily been put on hold, was reinstated: All the men of the X Corps were to resume their drive to the Yalu, with the goal of reaching the river as quickly as possible. This included Smith's Marines. The attack at Sudong was discounted now, an unfortunate but inconsequential speed bump on the way to certain victory. The idea of merely halting at the Chosin had been cast aside—Smith's Marines were to quickly pass by the reservoir and then keep on going to the Manchurian border. Almond assigned Smith a specific swath along the river, forty miles in length, that was to be his ultimate objective.

Once again, Smith and his staff were shocked by Almond's intemperateness. His plan, especially given this polar weather, seemed to them not merely ill-considered; it was crazy. When Smith met with Almond on November 14, he tried again to voice his doubts. Smith was still concerned primarily with how diffuse his division had become—his units were spread over far too many miles of terrain. Almond consented to let Smith concentrate his forces a bit but remained adamant about the haste with which they were to move toward the Yalu. "We've got to go barreling up that road," he said at one point. Upon hearing this, Smith blurted an involuntary response: "*No!*" Almond pretended not to hear him—otherwise he would have had to call Smith out for insubordination. The conference abruptly adjourned and Smith marched outside, hot as a hornet, saying to his staff, "We're not going anywhere until I get this division together."

For the first time, Smith was starting to see the situation whole: the terrain, the weather, the pressures from below and those from above, the enemy before him and the enemy within. As the pieces slowly came together, he could intuit the battlefield for what it would become—a perfect trap. It was starting to feel reminiscent, he thought, of Napoleon's disastrous 1812 campaign into Russia, in which the French Grande Armée pushed ever deeper into hostile country, distancing itself from its supply lines as winter steadily approached.

Smith recognized that he would be facing not one but four different foes: the mountains, the winter, the Chinese, and his own superiors. He would have to come up with ways to stall his division's advance without blatantly disobeying orders. He would have to figure out how to create depots along the route, and he would have to establish a central stronghold, a self-sustaining bastion in the wilderness. He knew he was not going to receive much help from Almond or X Corps. His Marines would have to rely upon their own wits and resources. *No one can save us but ourselves.*

What most concerned Smith was that his Marines were heading into a situation that, on the grand strategic scale, seemed conspicuously flawed. He recognized, of course, that matters of larger strategy were not properly his realm—he was only a field general—but he'd been in this place before, and he had seen the pointless carnage that could result from muddled strategy. Six years earlier, at the Battle of Peleliu, where Smith was assistant commander of the First Marine Division, the Marines suffered 6,525 casualties, including 1,252 dead, in a brutal two-month engagement that, in the end, served little purpose. The Marines had come to the tiny coral island to seize a Japanese airfield in support of MacArthur's planned invasion of Mindanao, in the Philippines. But then, in a late reversal, MacArthur decided to bypass Mindanao in favor of Leyte, thus obviating the need to take the Peleliu runway. The battle forged ahead anyway, and the airstrip was indeed taken, but it would play no significant role in any future campaigns.

Peleliu proved to be one of the costliest battles of the Pacific war, yet, tragically, the island could have been avoided altogether. Smith came to realize, said a biographer, "that all of that blood and sacrifice and pain had been for no strategic gain." Peleliu hung over him like a curse. He did not want to go down that road again. Losing Marines was one thing; losing them in the service of a dubious and ill-conceived strategy, he thought, bordered on the obscene.

On November 15, Smith sat down and wrote a long letter to General Clifton Cates, the commandant of the U.S. Marine Corps, in Washington. In a firm but even tone, he laid out the fix he was in and confessed his trepidations about the coming battle. At times, his words had a haunting quality, a note of "may this cup pass from me." But his letter proved to be profoundly prescient.

"Although the Chinese have withdrawn to the north," he said, "I have not pressed Litzenberg to make any rapid advance. . . . I do not like the prospect of stringing out a marine division along a single mountain road for 120 air miles from Hamhung to the [Manchurian] border." Smith told Cates that he had "little confidence in the tactical judgment of the [X] Corps or in the realism of their planning. . . . There is a continual splitting up of units and assignment of missions which puts them out on a limb. . . . Time and again I have tried to tell the Corps Commander that in a marine division he has a powerful instrument, but that it cannot help but lose its full effectiveness when dispersed."

Smith confessed that he did not know what to do next. "Someone in high authority will have to make up his mind as to what is our goal," he said. "My mission is still to advance to the border. I doubt the feasibility of supplying troops in this area during the winter or providing for the evacuation of the sick and wounded." Though Smith was prepared to carry out his mission, he saw what was coming and thought that what was being asked of his men was neither strategically necessary nor fair. "I believe a winter campaign in the mountains of Korea," he said, "is too much to ask of the American soldier or Marine."

"This letter may sound pessimistic," Smith concluded, "but it is not meant to be so. Our people are doing a creditable job, [and] their spirit is fine."

15

FATTENED FOR THE KILL

Hagaru

General Smith had found his spot, the place where, if it became necessary, he would build his fort and make his stand. It wasn't much to look at, just an opening in the mountains, a broad, flat place at the tip of an inlet of the Chosin Reservoir. Hagaru-ri, the village was called, though the Marines would shorten the name to "Hagaru." (The suffix, *ri*, meaning "town," seemed superfluous.) Hagaru looked, as one Marine account put it, "like a mining camp transplanted from the Klondike." It was a village of about five hundred souls—a snaggle of weather-scabbed shanties, mostly, though the place did boast a number of concrete buildings and one or two examples of what might euphemistically be called streets. The town had a little school and a church. Old telegraph wires danced on creosote poles. Oxen stamped the frozen mud, their heads wreathed in steam.

Around here, beside the many-fingered lake, the country offered little beneficence. The alkaline soil was stingy and full of stones. Gold had once been extracted from the surrounding mountains, but the old mine shafts were abandoned and boarded up. Most of the people around Hagaru lived in feudal poverty. On the town's edges were dead fields of millet and barley, rustling in the wind.

The lake was quite lovely, though; it had not entirely frozen over yet, and it gave off an inviting sheen. Lieutenant Joe Owen's mortar unit established a perimeter at the far north of the town, right on the shore. Owen remembered his first view of the lake. "It was sparkling blue against the thick pine forests that clothed the surrounding mountains. On that sunny day when we first saw it, the Chosin was a tranquil body of shimmering water, a scene of magnificent beauty."

General Smith had picked Hagaru as his forward base for several reasons. First, at least some infrastructure existed here on which he could build. As tiny and dismal as it was, Hagaru was the biggest town on the reservoir—the biggest town anywhere in this mountain domain. Then, too, Hagaru lay at a strategically important road juncture. Here, the main supply route forked. By veering right, one headed onto a road that wriggled up the east side of the reservoir; by veering left, one wound into the mountains for a ten-mile stretch before spilling onto a road that fed along the west side of the reservoir. Hagaru was thus the gateway that controlled all traffic along the lake.

But the most important reason Smith chose Hagaru was basic topography: It was the only place anywhere in the reservoir country that was wide and flat enough to accommodate an airstrip. Smith had surveyed the maps and had inspected the terrain by helicopter. Then, on November 16, he rode up to Hagaru in a heated station wagon with Major General Field Harris, commander of the First Marine Air Wing. Smith and Harris had wandered out to a crusty expanse on the southwest edge of town and found what Harris thought would be the ideal place to build an airstrip. The two men did some crude measuring, then squatted in the field and rooted around in the soil, inspecting its quality. It was dark and crumbly and did not seem to drain well—the kind of substrate that, in summer, would turn into a soupy bog. But in this weather, the dirt had frozen solid, and for that reason it would probably work fine. At last, Smith had something good to say about the extreme cold. "This," he quipped, "is one break we get from the winter."

The project Smith had in mind was not some piddling airstrip for Cessnas and spotter planes. (The town already had a crude runway that could accommodate small craft.) What the general wanted was an airfield long enough to handle C-47s and other transports. He planned to bring in enormous amounts of cargo—ammunition, food, medicine, fuel—and he wanted to turn this tiny place into some mountain approximation of O'Hare. He envisioned the big planes coming and going from dawn to dusk.

A few days earlier, when Smith had broached the idea of creating this airport in the middle of nowhere, Almond had failed to understand. Why was it necessary? Smith answered that, among other things, he imagined that Hagaru would become a triage hospital, with the wounded being brought in from the battlefields to the north. With a big airstrip, Smith could use large planes to haul out the casualties to hospitals on the coast or in Japan. An airstrip might save a lot of lives.

Almond was puzzled by this. "What casualties?" he said. He hadn't given the matter any thought.

=

In the end, Almond granted Smith permission to do whatever he pleased on the matter of an airstrip, so long as the Marines built it on their own. And so Smith enlisted his best engineer, Lieutenant Colonel John Partridge, the same man who had improvised the rafts that ferried Smith's heavy equipment across the Han River and later built the pontoon bridge that made crossing the Han a cinch for successive waves of X Corps units streaming in from Inchon. By November 20, Partridge's engineer battalion had made it up from Hamhung with a fleet of dozers and other earth-moving equipment, and now they got to work.

But Partridge found that he had a knotty problem: Hagaru sat at an elevation of four thousand feet, and for that altitude his manuals advised a minimum runway length of 3,900 feet to get the C-47s safely aloft—the thinner air required a longer strip. But according to his measurements, his engineers didn't quite have the available space to complete an airstrip of regulation length; a three-thousand-foot strip would have to do. As far as he knew, no aviator had ever tested these tight specs before. But Smith wanted him to take the risk.

So Partridge's explosives experts began to blast at the earth, and his machines scraped the frozen sod. This supposedly level spot turned out to be not so level after all—and the deeper they dug into it, the more they ran into shards and boulders, sometimes striking bedrock. The Caterpillar drivers toiled through the days and nights in the acrid halo of the floodlights, at times harassed by snipers' bullets. The machinery operators carried weapons—*every Marine a rifleman*—and were repeatedly forced to interrupt their work to fight off an enemy that most assuredly did not want to see an American airport built in this place. Partridge's men used pneumatic drills and jackhammers to break up the hardest places, and welders affixed steel teeth onto the bulldozer pans to bite deeper into the frozen soil.

Partridge's project had a quixotic quality: building an airfield where none should be, in a Siberian mountainscape, for a battle that the high command insisted would not happen, to placate the worries of a field general who believed otherwise and who had a sixth sense that big airplanes would be his salvation. It seemed grandly implausible, like Hannibal marching his elephants through the Alps. What made it more implausible was the tight deadline: Smith told Partridge he wanted the airstrip ready as soon as possible, but no later than December 1. And so, around the clock, the bulldozers kept grinding.

=

The airstrip was only one part of Smith's plan to make Hagaru the central node of the coming battle. The village would serve as his command post, his communications center, his field hospital. The heart of the division would rest within the safety of its stoutly defended perimeter. Here Smith would establish his biggest supply depots and ammo dumps. Here he would crank his howitzers into position to rain terror upon the distant hills. Hagaru would be like a protected cyst embedded in a foreign body, impervious to all assaults that might be mounted against it.

Litzenberg's Seventh Regiment had marched and motored the nearly ten miles up from Koto-ri and taken possession of Hagaru on November 15. Most of the way, they'd had sporadic contact with Chinese troops. But the enemy seemed mischievous and at times chimerical. If you pursued them, they vanished. If you ignored them, they reappeared. For days, they stayed just out of reach; they were, said one account, "invisible and yet seemed to be so many they were stirring the air itself." At times they caused real trouble, but mostly they pestered and harassed—playing, it seemed, an on-again, off-again game of cat and mouse. It was as though they wanted to remind the Marines that they were here, but at the same time wanted to encourage the Marines to keep on coming.

General Smith, who had made frequent trips up the road by jeep and by helicopter, was aware of this capricious behavior—and what he took to be the larger strategy behind it. Where it became most obvious to him was at an odd bottleneck called Funchilin Pass, a few miles west of Sudong.

The bridge at Funchilin Pass wasn't a typical bridge, not a span over a river or a stream. The structure was a bit more complicated than that. At this spot, a subterranean tunnel carrying large amounts of water from the Chosin Reservoir emerged from the mountain. The torrent fed into four large steel pipes—penstocks, they were called—that then redirected the water over a cliff and down a thousand-foot slope toward a series of power stations in the valley. A concrete bunker, housing a network of valves and spigots, was built into the sheer hillside. Next to it, a cantilevered concrete bridge passed over the four penstocks as they dropped into the chasm below. This improbable installation, perched so precariously on the mountain, was part of the larger Chosin hydroelectric complex.

What worried Smith about the Funchilin bridge was this curious fact: It had not been molested. If the Chinese had intended to halt or hinder Smith's advance, they would have blown it already. It was a gate governing entry into the reservoir high country, and it could easily be demolished. By leaving the bridge intact, they seemed to be ushering the Marines forward. Smith surmised that only after his entire division had passed safely over the bridge would the Chinese finally destroy it, thus sealing the Marines off from the safety of the coast. To Smith, it felt like a sly old trick, with shades of Moses parting the Red Sea waters just long enough to lure Pharaoh's soldiers to their drowning deaths.

On this point, as with so many others, Smith would prove prescient. Said Brigadier General Edward Craig, the division's assistant commander: "Smith was sure that they wanted us to come across, and that they were going to blow the bridge after we crossed, thus completely isolating us. It was shrewd of Smith to understand that." When Smith pointed out the perils of the Funchilin Pass bridge to General Almond, it did not register a response. Almond, said Craig, "seemed to have so little respect for the Chinese as fighting men, it was as if he didn't care."

≡

General Almond, in fairness, was a busy and distracted man during this particular week. He had been in nearly constant motion—in helicopters, on planes, in trucks and jeeps. The many units of his X Corps were spread over hundreds of square miles of terrain—Smith's Marine division was only one of them—so Almond had a lot to keep track of. Whatever his shortcomings, he could not be faulted for lack of energy or enthusiasm or, for that matter, courage. "As far as I could tell, he was utterly without fear," wrote Alexander Haig. "In combat situations, he exposed himself to mortal risks several times a day." Almond's habit of consciously placing himself in danger went back to his days in World War I in France, where, as a young soldier, he was wounded and won several decorations, including the Silver Star for heroism. Here in Korea, the hyperkinetic general would visit any forward command post and barge right into active combat zones.

Not that there was much action in the X Corps sector of North Korea. Aside from the Marines, few of Almond's units had reported significant contact with the enemy—either the North Koreans or the Chinese. Some of Almond's Army regiments were rushing for the Yalu, seemingly unimpeded. On the twenty-first of November, the leading battalions of the Army's Seventeenth Infantry reached the river at the North Korean town of Hyesanjin, and Almond flew in for the historic occasion. He stood at the Yalu's banks and posed for a triumphant picture with some of his commanders. Across the swirling waters, a few Chinese sentries could be seen, calmly going about their errands. "They made no effort to fire on us," recalled Haig, who was traveling with Almond's entourage. "Their breath vaporized in the frigid air."

Almond got General MacArthur on the radio and told him the

news. "Heartiest congratulations, Ned," MacArthur's voice squawked. He told Almond that he and his men had "hit the jackpot." Almond and some of the others enthusiastically whipped out their equipment for a ritual piss into the stout waters of the Yalu. If this seemed to be flirting with fate, or conduct unbecoming of a general, Almond couldn't resist the urge. General George Patton, whom Almond admired and viewed as a model, had urinated in the Rhine when his Third Army had reached that broad river; Almond felt compelled to do the same here.

Almond beamed with pride at what his men had accomplished— never mind that most of them had never seen the enemy. The drive to the Yalu, he believed, would go down in the annals of heroic martial feats. "The fact that only twenty days ago," Almond wrote the next day, "this division landed amphibiously . . . and advanced 200 miles over tortuous mountain terrain and fought successfully against a determined foe in subzero weather will be recorded in history as an outstanding military achievement."

A few days later, another unit from Almond's X Corps, Task Force Kingston, reached the Yalu at a place called Singalpajin, where they engaged in a brutal house-to-house fight, rooting out the remnant North Korean enemy. Then they, too, urinated in the Yalu—it was becoming a fad. Others filled bottles with river water to keep as souvenirs. The men of Task Force Kingston were the second group of American troops to make it to the brink of Manchuria; they would also be the last.

Along the way, they suffered a bizarre and unexpected casualty that was taken by some as a bad omen: One of their soldiers was mauled and killed by a Siberian tiger.

≡

Thanksgiving arrived on November 23. General Smith had hoped to spend it at his command post with his immediate staff. Admiral Doyle had sent him a whole cooked turkey from the galley of one of the Navy ships docked in the harbor. But then Smith got a last-minute invitation to a formal dinner at X Corps headquarters with

General Almond and a brace of high-ranking officers. It was about the last place Smith wanted to spend Thanksgiving, and the last person he wanted to spend it with. But he knew that declining Almond's invitation would have been supremely bad form.

Almond, by this point, had truly made himself at home in Hamhung. He liked his creature comforts and made sure they were close at hand. His aide, Alexander Haig, had been put in charge of an effort to requisition a private home to serve as Almond's residence. Haig had found a lovely Japanese-style villa and, after being given a budget to have it redecorated to Almond's tastes, hired Korean artisans to build, among other things, a sunken bathtub set in mosaic tiles. Haig had also taken it upon himself to grace the villa's reception halls with hand-painted vases that he'd bought in a Hamhung bazaar.

One day, South Korean president Syngman Rhee and his wife visited Hamhung to give a series of reassuring talks to the North Koreans, and while there, they stopped by to visit Almond's new residence. Haig's elegant vases were arranged with beautiful fresh flowers. Almond, pointing them out to Mrs. Rhee, gave Haig the credit for having located such beautiful objets d'art in a war-ravaged city. Wrote Haig, "Mrs. Rhee, a brusque and outspoken woman, fixed the general with a disdainful stare. 'Then your aide should know for future reference,' she said, 'that in Korea vases of this kind are used as chamber pots.'"

When General Smith arrived at Almond's headquarters, he found a lavish spread for a total of twenty-eight guests. The dining room featured a cocktail bar, white linen tablecloths and napkins, fine china, polished silverware, even place cards at each seat. The best delicacies had been flown in from Tokyo, compliments of MacArthur. "It was a plush state of affairs," Alpha Bowser recalled, with "all the appointments one would expect in, say, a formal function in Washington."

Smith's own tastes tended toward the Spartan—he was "as spare as the Marine Corps itself," wrote one Marine historian. He felt a bit queasy about eating such a meal, with such splendid appointments, knowing that his regiments were freezing out in the field, and in harm's way. The extravagance of the feast felt unseemly. Once again, he thought, Almond was tone-deaf to the true battlefront situation.

To Smith, this Thanksgiving dinner had taken on the quality of a fete that anticipated the war's end.

Almond, for his part, seemed to regard Smith as a wet blanket. They had so much to be thankful for, so much to celebrate. After all, Almond had gazed upon the waters of the Yalu only two days before. He had seen for himself that the talisman was within reach. Why couldn't Smith loosen up and enjoy himself?

Smith's Marines—indeed, all the men of X Corps—had been treated to their own version of a turkey dinner, albeit a less elaborate one. Across the battlefield, homesick young American men had enjoyed a reasonable facsimile of a holiday feast. The menu included shrimp cocktail, stuffed olives, roast young tom turkey with cranberry sauce, candied sweet potatoes, and mincemeat pie. How the cooks and the logistics people had pulled it off, no one could guess—the exertion and expense had to have been stupendous. "High in the bitter land, Americans ate Thanksgiving dinner," wrote historian T. R. Fehrenbach. "Depending on their tactical position, they ate well or plainly—but most received turkey and all the trimmings, brought into this savage country at great effort."

The hungry men of Litzenberg's Seventh Regiment, far up in Hagaru, were about as far from the distribution points as anyone. When the raw turkeys reached Hagaru, they were frozen rock solid, and the cooks had to figure out how to thaw them. Finally a solution was devised: The kitchen details piled the cold birds into a small mountain atop a pair of burning stoves, then covered the mass with tent canvas weighted down with snow. The frozen fowl baked in this makeshift sauna through the night, and by morning they were sufficiently thawed for the cooks to roast.

Most of the men in the field savored this turkey dinner as they had no other. They would talk about it the rest of their lives: A Thanksgiving feast on the far side of the planet, in the snowy shadow of Manchuria. But the awareness that the Chinese were nearby crept into their thoughts. Perhaps it was gallows humor, but as they tucked into their plates of food, an unsettling thought did occur to more than a few of the Marines. As one of Litzenberg's men put it: "We kind of wondered: Were we just being fattened up for the kill?"

16

NEVER TOO LATE TO TALK

New York City

At a little past six o'clock on the morning of November 24—the day after Thanksgiving—the Chinese delegates marched off the plane and onto the tarmac at New York's Idlewild Airport. No emissaries from the People's Republic of China had ever visited the United States before, and now here were nine—seven men, two women—who'd come to speak for Mao's fledgling government at a special session to be held at the United Nations. They had flown across Mongolia, across the Soviet Union and Europe, then on to London, where they had caught this British Overseas Airways flight to America. They were ambassadors from the world's most populous nation, but, because Truman had refused to recognize the Communist regime as China's legitimate government, they walked onto American soil as aliens, denizens of a country with no standing—and a country with whom America was now on the precipice of full-scale war. Their modest welcoming party included a protocol officer from the U.N., several Russian dignitaries, and a pack of journalists. Not a single United States official deigned, or dared, to show up.

The leader of the delegation, Ambassador Wu Xiuquan, emerged from the plane to sniff the air of what he called the world's "most arrogant imperialist state." Wu was a contentious, sinewy man with a high-pitched voice, whose face had been scarred by a bullet during the Chinese Civil War. Educated in the Soviet Union, forty-two years old, he was a trusted colleague of Zhou Enlai, Mao's foreign minister. Wu was understandably wary. He had come to the lion's den, to the coruscating metropolis that was at the beating heart of American capitalism, to lodge an official complaint with the United Nations Security Council about the U.S. naval presence in the Strait of Taiwan and its rapid drive in Korea toward the Yalu River.

"A heavy responsibility had been placed on our shoulders," Wu later wrote. "We had moved from the military battlefield to the debating platform to wage a tit-for-tat struggle." It would be, said Wu, a "face-to-face struggle against the number-one imperialist state on its own turf."

Although Wu insisted that the United States had committed an act of "armed aggression" against China, he met reporters at the airport with a salutation to all Americans. "A profound friendship has always existed between the Chinese people and the American people," Wu said. "I wish to avail myself of this opportunity to convey my greetings to peace-loving people in the United States."

Wu and his fellow delegates were driven to Manhattan, where they checked into a block of nine adjoining rooms on the ninth floor of the Waldorf-Astoria. Though they were impressed by the opulent appointments, they balked at the exorbitant rates and found the room service fare nearly inedible. Wu and his comrades secured a stash of hot Mexican chiles to enliven the nauseatingly bland dishes.

A detail of New York police guarded their rooms around the clock—their U.N. hosts feared that perhaps some overly patriotic vigilante, or some embittered gold-star parent, might try to break into their rooms and cause an international incident. The Chinese were certain their quarters had been bugged. They kept a radio blaring in their office suite day and night, Wu said, to create a "constant din that might interfere with attempts to eavesdrop." Whenever they had sensitive material to discuss, they met in a nearby park.

Everyone at the U.N. seemed aware of the official purpose of the Chinese visit—to protest American doings in Korea and Taiwan—but in back of that awareness floated an ardent wish that the parties might still negotiate a settlement before the world slipped into full-blown war. Though the Chinese were putting up a bellicose front, a number of optimists clung to the hope that perhaps Mao's emissaries had secretly come to negotiate an agreement. After so many years of civil war, the Chinese surely could not afford a conflict with the United States and her allies.

It was widely reported that the American, British, and French governments had reached a tentative agreement to propose a ceasefire and a buffer zone along the Yalu. This no-man's-land could include all the North Korea–based hydroelectric plants that supplied power to Manchuria, as this was thought to be one of the major points of Chinese contention. Throughout the U.N., there was hope that a settlement could be worked out. The Chinese presence here in New York, many felt, was the world's last chance for sanity to prevail. One Far East delegate expressed the attitude this way: "It is never too late to talk; it is always too early to fight."

In truth, a shroud of mystery hung over the Chinese visit. Why had they come so far, with such a large entourage? If all they wished to do was issue a complaint, they could have done it by cable. Their sojourn to New York would seem to be an unnecessary and expensive bit of diplomatic theater. Trying to divine their true purpose in New York, said one U.N. delegate, was "like flying blind through an uncharted mountain range."

The nine Chinese delegates, holed up in their posh rooms at the Waldorf and guarded around the clock, waited for their chance to address the U.N.

=

The same day the Chinese delegation arrived in New York, an unusually powerful storm system began to form across the eastern third of the United States, one that would temporarily divert the attention of the Truman administration from the looming conflict with China. The storm started with an Arctic cold front that fingered down through Ohio and eastern Kentucky. Across Appalachia, the mercury dropped from the fifties to the teens within a few hours. By the next day, as the cold air mass barreled toward the east, a vast pocket of warm, wet Atlantic air from the Carolinas began to wrap underneath it. The storm had become an "extratropical cyclone." Huge amounts of snow began to fall across Ohio, Kentucky, and Pennsylvania. In one town, deep in the hollows of West Virginia, fifty-seven inches of snow fell in just over a day.

On the east side of the front, gale-force winds began to buffet New York and New England, cutting electricity from more than a million households. Manhattan recorded a peak gust of nearly a hundred miles per hour, and surging seas breached the dikes at LaGuardia Airport, flooding the runways. The nasty weather forced the Chinese delegates to stay inside their Waldorf-Astoria rooms for two days.

The event, which continued to rage through Thanksgiving weekend and beyond, would affect twenty-two states and would kill 353 people. On some of the worst-hit highways, National Guardsmen were brought in to remove snow with tanks and flamethrowers. Newspapers called it the Storm of the Century. Whatever it was, the cyclone was an anomaly that would be studied for decades. "The Great Appalachian Storm of 1950," as it would officially become known, was the costliest and most destructive storm then recorded in U.S. history— a wintry vortex that few saw coming, and few understood even after it had arrived.

17

NEVER A MORE DARING FLIGHT

Over the Yalu

The day after Thanksgiving, Douglas MacArthur decided it was time to set foot on the battlefield and breathe the air of North Korea, time for the world to be reminded that he was the architect behind momentous events that were about to transpire. So the supreme commander gathered his favorite journalists and his favorite aides and brought the entourage over from Tokyo aboard his favorite Air Force plane, the *Bataan*. A few hours later, the four-propeller Lockheed Constellation touched down at Eighth Army headquarters on the Chongchon River—over on the west side of the peninsula.

MacArthur stepped out into the bitter weather wearing a sprightly checkered scarf and, his thespian sensibilities fully aroused, sent a message to the troops in which he laid out his plan for the final offensive. In a communiqué he sent to the troops, MacArthur spoke again of pincers and vises and other dashing movements. The United Nations was initiating what he called a "massive compression envelopment." The remaining terrain of North Korea had been organized into "sectors" in which the enemy would be systematically "interdicted" by air. By his description, what was about to be unleashed upon the enemy in this final phase sounded like some intricate convergence of military science and the occult.

Perhaps even stranger than MacArthur's baroque language was the extent to which he felt compelled to reveal his strategy: He was giving away his battle plan to the world, including the enemy. He was so confident in his success that he saw no reason for secrets. It was a bravura performance, even for him—the martial equivalent of Babe Ruth at the plate, pointing out the spot in the center-field bleachers where the homer was going to land.

MacArthur assured the troops that this final maneuver would "for all practical purposes end the war." The Eighth Army would seize the western half of the border, while Almond's X Corps would claim the eastern half. Hostilities, MacArthur predicted, would end in a matter of weeks. Then he made a pronouncement that would become infamous (though he later insisted he'd been misquoted): The boys, he said, would be home by Christmas.

MacArthur announced that a newly intensified air campaign was already causing devastation across much of North Korea and inflicting massive casualties on the enemy. Though MacArthur didn't know it, these aerial attacks were inflicting considerable damage on Chinese troops, as well. (In fact, the very next day, a U.N. airstrike over North Pyongan Province, North Korea, would yield a prominent Chinese casualty: Mao Anying, Chairman Mao's eldest son, was incinerated by a napalm bomb. He was twenty-eight years old.)

MacArthur met briefly with General Walton Walker's officers, then wished his armies Godspeed and waved them off. Though he had spent only a few hours on the ground, he was ready to leave. It had been time enough for the newsmen to snap their shots and gather some quotes. MacArthur saw no point in lingering. What mattered was that the "compression envelopment" had commenced—though that phrase was too much of a mouthful for the journalists, who had already concocted neater descriptions for it. The Home for Christmas campaign, they called it. The end-the-war offensive.

After much vigorous saluting on the tarmac, the *Bataan* roared off. Once in the air, MacArthur did something odd: He instructed the Air Force pilot, Lieutenant Colonel Tony Story, to alter the flight plan. He didn't want to head straight for Japan; he wanted to detour and fly over the length of the Yalu, to see the river with his own eyes. He wanted to peer down at the Manchurian border and hunt for signs of the much discussed but seldom seen Chinese.

Story was alarmed and puzzled by the request—so were the members of MacArthur's staff. It seemed to them a gratuitously dangerous thing to do. The Russians, or the Chinese, might perceive such a flight path, skirting the international border, as a provocative act. Hewing to the bends of the river, the defenseless transport plane would stay in nearly constant reach of enemy antiaircraft batteries. MacArthur appeared to be tempting fate.

His staff tried to dissuade him, but MacArthur brushed them off, insisting that (shades of Inchon) the "very audacity of the flight would be its own protection." One officer discreetly pleaded with Story to find some other North Korean river to fly over. MacArthur couldn't possibly know the difference, he whispered. But Story demurred, saying, "I couldn't lie to the chief." Then his close adviser General Courtney Whitney suggested that, at the least, they should take the precaution of strapping on parachutes. MacArthur could only chuckle at these fuddy-duddies. "You gentlemen can wear them if you want to," he said, "but I'll stick with the plane."

Story decided that it would be prudent to radio for a fighter escort. A couple of jets rushed over from Kimpo Field and caught up to the *Bataan*. Though Story felt safer, the presence of fighters circling above only made MacArthur's flyover seem more provocative.

They headed north to the mouth of the Yalu, then turned east and followed the meandering course of the river at an altitude of five thousand feet. "At this height," MacArthur wrote, "we could observe in detail the entire area of international No-Man's-Land all the way to the Siberian border. All that spread before our eyes was an endless expanse of utterly barren countryside, jagged hills, yawning crevices, and the black waters of the Yalu locked in the silent death grip of snow and ice."

=

When it came to the art of aerial reconnaissance, MacArthur had no particular expertise—though he seemed to think he did. As he later put it, he believed that merely by flying over the Yalu, he could "interpret with my own long experience what was going on behind the enemy's lines." But gazing into the sprawling wilderness, MacArthur could detect no trace of the Chinese: no movements, no supply depots, no columns of stick figures beetling over the ground. All he saw was "a merciless wasteland."

Yet the Chinese were there, hundreds of thousands of them. They were masters of camouflage. Some had waded across the icy Yalu, while others had shuffled over prefabricated pontoon bridges that Chinese engineers had submerged six inches below the surface. Still others stripped down and swam the freezing waters, emerging, according to one Chinese historian, "with ice hanging on their bodies, like gods in silver armor."

Mao's army, wrote the combat historian S. L. A. Marshall, was a "phantom which casts no shadow. Its main secrets—its strength, its position, and its initiative—had been kept to perfection, and therefore it was doubly armed." More than a quarter million Chinese soldiers had crossed into North Korea, while another half million were massed north of the border in Manchuria. Now they were waiting. Mao knew that the farther the Americans advanced toward the Yalu, the more stretched out they would become—their supply lines overextended, their communications interrupted, their lines too thin for defense.

Mao and Peng had decided to send one of their best armies, the Ninth Army Group, to the Chosin Reservoir to confront the First Marine Division, which they believed to be the strongest of all the U.N. forces approaching the Manchurian border. "It is said," Mao wrote to Peng, "that the 1st Marine Division has the highest combat effectiveness in the American armed forces. Your generals should make its destruction their main effort." The Chinese officers had passed down through the ranks the notion that the U.S. Marines were uniquely diabolical—bloodthirsty murderers and rapists and, according to one account, "highly competent criminals." A rumor was passed around that to be inducted into the Corps, a prospective Marine had to kill a member of his own family; a distant cousin wouldn't suffice—it had to be a direct relative. Another rumor said Marines were known to eat babies.

Now the Ninth Army Group, under the command of General Song Shi-lun, began to move into the mountains around the reservoir. The slopes were so steep and slick that many of Song's shaggy ponies, burdened with ammunition, had refused to climb them. But

the Chinese had improvised a solution: They laid down their sleeping mats and blankets along the path of the march, to give the ani-
mals traction. These men, most of them veterans of China's civil war,
were seasoned peasant fighters, adept at marching long distances on
little food—often nothing more than a few balls of sorghum, millet,
coarse rice, dried peas, and sesame. They carried primitive weapons,
but they had the advantage of a zealous indoctrination—they'd been
taught to believe that the northward-marching Americans posed a
true threat to the young People's Republic. America seemed to stand
in direct opposition to the revolution that Mao's armies had fought so
tenaciously to bring to fruition. Which was to say, many of the men
of the Ninth Army Group had good reasons to hate the Americans.

Now the armies were well established in the highlands around the
reservoir. On the eve of battle, at his headquarters not far from a tiny
lakeside village called Yudam-ni, General Song told his troops, "Soon
we will meet the American Marines in battle, and we will destroy
them. When they are defeated, the enemy army will collapse and our
country will be free from the threat of aggression."

Then Song exhorted his men: "I want you to kill the Marines as
you would snakes in your home."

≡

The *Bataan* droned along the borderland. With each bend in the
river, MacArthur's mood brightened. The Chinese were nowhere to
be seen. And in his peculiarly self-referential mind, his not seeing
them was proof that they weren't there.

Much about the reconnaissance flight captured the idiosyncra-
sies of MacArthur's personality: It was brash, extravagant, and unexpected. It had the feel of a stirring adventure, with an element of
real (if also manufactured) risk. But in back of it was an undertow of
vulnerability, a gnawing need to know what was down there, to size
up his foe. He was a man who believed in destiny, in consulting the
auguries. His aides would later praise him for his bravery—his chief
intelligence officer, Charles Willoughby, would go so far as to say that
"the air has never seen a more daring flight." MacArthur would even

earn a medal of valor for this dubious mission: The Air Force later awarded him the Distinguished Flying Cross and the honorary wings of a combat pilot.

Still, the Yalu detour had accomplished nothing except to harden MacArthur's belief in the rightness of a course that was madly wrong. This was his moment of no return. By now, said one prominent Army general, MacArthur had come to resemble "a Greek hero of old, marching to an unkind and inexorable fate."

The supreme commander puffed his pipe and bantered with his aides while still staring out the window. Upon reaching the eastern edge of the Korean Peninsula, the *Bataan* banked south over the sea and headed for Japan.

BOOK THREE

THE

RESERVOIR

The enemy advances, we retreat; the enemy camps, we harass; the enemy tires, we attack; the enemy retreats, we pursue.

—MAO ZEDONG

18

EASY FOR US, TOUGH FOR OTHERS

Yudam-ni

Yudam-ni was a mountain hamlet near the shores of the Chosin Reservoir, a speck of a place, a crossroads. Just to the east, in swirls of mist, a cove of lake ice gleamed the color of nickel. Five steep ridges, splotched with stunted pines, hemmed in the town and inspired among the Marines a nagging feeling that they were being watched. In the waning light, on the valley floor, a dozen mud-and-thatch huts, mostly vacant, threw shadows over the stubbly fields. A few villagers cowered in doorways and smiled awkwardly. They feared the American strangers and their machines—and nursed a thinning hope that the war would quickly move on.

Something momentous was happening in the mountains around Yudam-ni, and the villagers knew it. Plays of shadows. Human voices murmuring through the hollows. Deer bolting down from the ridges in large numbers, as though spooked. Everyone knew the Chinese were gathering somewhere to the north. It was only a question of how numerous their forces had become, and what their true intentions might be.

By the afternoon of November 27, more than eight thousand Marines had arrived in Yudam-ni, and the place was an industrial hive. The Marines had quickly transformed the valley into a sprawl of equipment depots, mess tents, trash heaps, munitions dumps. A layer of exhaust hung over the town like a grimy halo, and grids of canvas shelters billowed in the bracing wind. Men hunched around barrels, warming their hands over sparky fires. Occasionally a plane would appear and a mystery package, attached to a parachute, would drop from the sky.

The Marines were using Yudam-ni as a temporary staging area from which to launch the drive across the Taebaek Mountains. From here, they were supposed to march fifty miles toward the west to shore up the flank of General Walker's Eighth Army, then turn north for the final push to the Yalu River. Korea would be unified. The war would be over in a week or two—that was the word coming out of Tokyo.

But the optimism radiating from MacArthur's headquarters seemed sharply at odds with the mood around Yudam-ni. Here, worry lined the faces of the men. High along the ridges, Marine units were digging in for the night, using their entrenching tools, and sometimes explosives, to gouge shallow foxholes from the frozen earth. Individual platoons tightened their perimeters, while wiremen scurried up the hills, unspooling wheels of communications coil.

The regiments at Yudam-ni—Litzenberg's Seventh and Lieutenant Colonel Raymond Murray's Fifth, along with elements from the Eleventh, a heavy artillery regiment—understood that they were the tip of the division spear. They had advanced the farthest north, the farthest west, higher and deeper into the mountains than any of Smith's forces. As the sun sank toward the ridgeline, the men knew how far removed they were from help. Exposed as they were out here, they felt they'd become, said one account, "the plaything of the old men who directed them, the old men who were always fighting the last war."

≡

One of the outfits digging in for the night was the Seventh Marines' Company E, also known as "Easy." The 170 men of this scrappy rifle company—"Easy for us, tough for others" was their slogan—had taken up a position on North Ridge, an imposing escarpment that overlooked Yudam-ni. Specifically, Easy was supposed to defend a spur along the ridge called Hill 1282. (The number signified its height in meters, as noted on the old Japanese topographical maps the Marines were using.) Three platoons of Easy Company had begun to arrange themselves in a semicircle near the summit of the hill. At their backs, some seven hundred feet downslope, was Yudam-ni and their fellow Marines; in front of them was another steep slope, spreading out below. Easy's job was to keep a close watch on that slope, to make sure the CCF didn't come marching up it in the dead of night. The fear was that the Chinese might try to pour over North Ridge and overrun the Marine command posts and medical tents in the village itself.

Anchoring the center of Hill 1282, facing toward the northeast, was Easy's Second Platoon, led by a legendary Marine from Arkansas named John Yancey. First Lieutenant Yancey, thirty-two years old, was a World War II veteran who had won a Navy Cross at Guadalcanal and had also fought at Saipan. His ruddy face was roped with scars. On Guadalcanal, he had lived behind enemy lines for a month, subsisting on nothing but rice. As was then the grim custom among Marines fighting a fanatical enemy, he had collected a number of souvenirs from his bouts of hand-to-hand combat—including two Japanese pistols, a bayonet, and a gold tooth that, in his youthful battle ardor, he was said to have extracted from the corpse of one of his most hated foes after a horrific fight. Yancey, according to one account, had "learned his own lessons in a hard school, the hardest there was." In a single action, he had killed thirty-six Japanese soldiers—"before breakfast," his citation noted. Among those casualties was an officer who had, as Yancey put it, "attempted with great vigor to decapitate me with a sword."

After the war, Yancey had been a lineman for the Razorbacks of the University of Arkansas. Now he owned a spirits shop in Little Rock and also a nightclub, called the Gung-Ho. He had recently become a father: The very day he came ashore at Inchon, his wife, JoAnn, gave birth to a baby girl. Yancey's punctilious posture, suave smile, and prim black mustache gave him the air of a maître d'—all that was missing was the white jacket. He was a perfectionist, that was certain, and he could be hard on "the kids," as he called his platoon members. He cursed in torrents and constantly issued commands in strident barks.

Still, the men of Second Platoon adored Yancey. He was a bona fide war hero, someone who came alive in combat and had a special talent for it. People said he was indestructible. The strut in his step was infectious. "Yancey was the kind of person I'd read about but never thought I would meet in real life," said Private James Gallagher, a machine gunner from Philadelphia. "He let us know early on that he would give the orders and we would follow him." Although he was from the South, he was quick to criticize any of his men who showed even a hint of discrimination toward black Marines in the newly desegregated Corps. Yancey was, said Easy Company's Ray Walker, "one of those natural born troop leaders, possessing both the charisma and the steel nerve."

"When Yancey was with us," said another member of the platoon, the men "had a kind of Valhalla complex." Speaking in his grumbly baritone, Yancey loved to quote—from memory—long verses of Kipling, his favorite poet, or certain lines from O. Henry, his favorite short-story writer. Uncannily calm in battle, he would utter swashbuckling exhortations that seemed straight out of a John Wayne movie. "Here they come, boys!" his men heard him yell during one firefight at Sudong. "Stand fast and die like Marines!" War, he said, quoting John Stuart Mill, was "an ugly thing, but not the ugliest of things. . . . A man who has nothing which he is willing to fight for, nothing which he cares more about than he does about his personal safety, is a miserable creature, who has no chance of being free."

But Lieutenant Yancey's personality had a whimsical aspect, too. When the work was done, when the danger had passed, he could be a mischief maker—and a master of the fine southern art of bullshitting. He was a "country fella" who'd spent time in the Ozarks, noted one Navy corpsman who served under him. "His banter was always upbeat. He made those youngsters forget they needed a hot meal, that they were cold and scared and homesick."

The stories that surrounded Yancey were almost criminally colorful, and many of them were true. On an abandoned block of Uijeongbu, one of Seoul's suburbs, he had used composition C explosives

General Oliver Prince Smith,
photographed during World War II
Marine Corps History Division

(RIGHT) General Edward Almond,
commander of X Corps   *MacArthur Memorial
Library*

(BELOW) General Douglas MacArthur
(seated) and General Almond (right)
follow the progress of the Inchon
invasion from the deck of the USS
Mount McKinley.   *MacArthur Memorial Library*

Units of the First Marine Division, spearhead of the invasion, ascend the Inchon seawall. *Marine Corps History Division*

The Inchon beachhead expands one day after the amphibious landings. *National Archives*

General MacArthur congratulates
General Smith on the success of the
Inchon landings. *MacArthur Memorial Library*

(LEFT) Marines, carrying
a wounded comrade,
encounter sniper fire in the
heart of Seoul. *Marine Corps History Division*

(BELOW) U.N. troops round
up North Korean prisoners
in downtown Seoul. *Naval Historical Center*

Lee Bae-suk as a young man in Korea *Courtesy of the Lee family*

President Truman and General MacArthur meet on the tarmac at Wake Island just before dawn. *MacArthur Memorial Library*

General Song Shi-lun, commander of the Ninth Army Group of the Chinese Communist forces *Marine Corps History Division*

At the cry of a bugle, Chinese soldiers mount an attack. *Army Heritage and Education Center*

Chinese prisoners, captured at Sudong, await
interrogation. *Marine Corps History Division*

The Marines find a propaganda poster in
an enemy bunker, en route to the Chosin
Reservoir. *Marine Corps History Division*

(ABOVE) General MacArthur scours the terrain during a reconnaissance flight over the Yalu River in late November 1950. *MacArthur Memorial Library*

(BELOW) Colonel Homer Litzenberg, commander of the First Marine Division's Seventh Regiment *Marine Corps History Division*

(BELOW) General Charles Willoughby, MacArthur's chief of intelligence *MacArthur Memorial Library*

(ABOVE) Lieutenant Colonel Ray Murray, commander of the Fifth Regiment. *Marine Corps History Division*

(BELOW) Colonel Lewis "Chesty" Puller, commander of the First Regiment *Marine Corps History Division*

(ABOVE) Lieutenant John Yancey of Easy Company *MacArthur Museum of Arkansas Military History*

(BELOW) Captain William Barber, commander of Fox Company *Marine Corps History Division*

(ABOVE) Private Hector Cafferata of Fox Company *Marine Corps History Division*

The Chosin Reservoir in winter   *National Archives*

Aerial view of the village of Hagaru-ri. The airstrip, newly carved from the frozen ground, is visible in the center of the image.   *Marine Corps History Division*

to blast open the vault of what appeared to be an abandoned bank. When the dust settled, he found a great quantity of North Korean occupation notes stacked in bundles. The war scrip was apparently valueless—but no matter. He had some of his men load the currency into a Russian jeep they'd captured and later distributed it throughout the company. Per his orders, everyone was given at least one bundle of cash. Several days later, Yancey returned from a beer-foraging trip to his unit's bivouac site. To his dismay, things had changed drastically during his brief absence. A succulent pig was roasting, booze had been procured, Korean girls frolicked everywhere about the place, and a brothel had been established in a back room, "from which came squeals of delight."

Yancey, fuming mad, summoned his platoon sergeant and lit into him for letting things go to hell. *Prostitutes?* "But the men are paying them with the money you stole from the bank at Uijeongbu," the sergeant replied. "These people think we're just a bunch of rich, crazy Americans."

Yancey had no choice but to let it go.

≡

At dusk, Yancey patrolled the perimeter and inspected the two-man foxholes he'd had his platoon digging all afternoon along Hill 1282. The men were sweaty and sore and exhausted from the toil. This ground was impossible, like chipping at solid concrete. Their entrenching tools merely skipped off the flinty ground with a hollow clang. But Yancey reminded them that their lives depended on depth: Every inch counted. "A little more, kids!" he kept growling at them. "A little deeper now!"

Yancey had them reinforce their foxholes with barricades of brush and rocks. He set up two men with .30 caliber machine guns at opposite ends, to protect Second Platoon's flanks. He surveyed the amphitheater of hills below him and tried to read the angles, tried to see it from the Chinese vantage point. Where were the weak points? Where would his men be most vulnerable to mortar concentrations? He sent some Marines out to arrange a circuit of trip flares, while others filled ration tins with pebbles and strung up these crude rattles to long runs of communications wire.

About forty yards behind Yancey's position, set in a declivity behind some boulders, was the Easy Company command post, where Yancey's superior, Captain Walter Phillips, was also getting set up for the night. Yancey went down to confer with Phillips and his company executive, First Lieutenant Raymond Ball. Behind the CP, mortar specialists were digging into the snow, and Company E's radioman was testing his frequencies. Yancey and Phillips talked awhile in the warming tent that had been set up. They wished each other good luck, and Yancey returned to the top for the night.

Back with his men, Yancey gave one of his little spiels on the importance of foot care. He was obsessed with the topic and was constantly reminding his charges to change their socks, rub their feet, wiggle their toes, and slather on generous daubs of boric-acid ointment. In the Pacific, he had seen the horrors of "jungle rot" and other podiatric ailments, but this was far worse. Frostbite was a silent and insidious enemy, he said, and every bit as lethal as the Chinese. But if you got religion and followed Yancey's regimen, you at least had a chance, he said, to come out of these mountains with feet instead of stumps.

Throughout the day, Yancey had been noticing that Private Stanley Robinson had been limping around like a cripple, wincing in pain. Robinson was a tall, scrawny, obstreperous kid who seemed to have a nose for trouble. Yancey liked him but considered him the "platoon delinquent." He instructed Robinson to remove his boots. The sight was sickening. The early signs of frostbite—swelling, blood-filled blisters, white scales on red skin—were obvious. The skin had sloughed off of both anklebones. A medical corpsman took one look at Robinson's feet and declared them beyond his resources to treat.

Yancey ordered Robinson to turn in his rifle and report to the infirmary in Yudam-ni. "You're going down the hill."

"Hell I am," Robinson protested. Defiant though he was, he looked up to Yancey like a father, and felt most at home in the field, under his watch.

But Yancey would have none of it. "Don't fuck with me, Robbie."

And so Robinson handed over his Browning Automatic Rifle. Still sulking, he shambled down the slick hill.

When the sun disappeared behind Sakkat Mountain, to the west, Hill 1282 was almost immediately plunged into darkness. Yancey kept bundles of kindling strapped to his pack, and soon he had a fire going in a protected place near the top. In the flickering shadows, he issued an order that gave the platoon pause. Down the line, the word spread: "Mr. Yancey wants to see bayonets on the ends of those rifles."

From their foxholes, the men of Easy Company's Second Platoon murmured uncomfortably. *What's Yancey know that we don't?*

≡

A little after 6:00 p.m., a bright gibbous moon, four days past full, peeked over the southeastern horizon to light up the slopes of North Ridge. It was a beautiful, sharp night. The ground fog was clearing. Yancey could see the reservoir: a white expanse with black bald spots where the wind had swept the snow off the ice. The valley was calm. The temperature had dropped to twenty below zero.

In the moonlight, Yancey could make out the jumbled ridges where pockets of other Marines were also settling in for the night. At nearly every point on the compass, another group of defenders occupied some godforsaken scrap of high ground: Hill 1203, Hill 1426, Hill 1294, Hill 1276, Hill 1240. Each unit, on each promontory, would have to tend to its own safety. But the men of Easy Company could at least take comfort in knowing they weren't entirely alone, that the travails of their shivering vigil were shared by some poor Marine bastards across the way.

Yancey realized that the moon was rising behind his position, an infelicitous angle that was throwing his men into silhouette and thus providing clearer targets for any enemy approaching from below. He put the platoon on 50 percent alert—that is, in each foxhole, one man would try to sleep while his buddy kept his eyes peeled, rifle at the ready. A few quiet hours passed, but then, at 9:45, the Easy Company radioman picked up some bad news from Dog Company, which was situated about a thousand yards away on an adjoining hill. "Heads up over there!" came the warning from Dog. "One of our guys just got bayoneted in his bag."

A few minutes after that, Yancey could faintly discern white shadows flitting over the ground, moving toward the crown of 1282. Then he saw the livid flashes of the Chinese burp guns as bullets peppered the hillside. This wasn't much of an attack, Yancey thought—quite tepid, in fact. He surmised that it was intended merely as a probing action, and he instructed his machine gunners to hold fire at first, so as not to reveal their positions. The Chinese were only testing, trying to locate the salient points and weak links in Yancey's line. Still, a vigorous firefight broke out for a few minutes. Yancey stalked the periphery, exhorting his men.

Gallagher opened up with his .30 caliber machine gun and mowed down a file of Chinese who were making straight for his position. They wore white quilted coats, fur-lined hats, and canvas shoes that looked like sneakers. Many of them sported white capes. Gallagher dropped one enemy soldier at point-blank range; the attacker fell with an elbow touching one of the tripod legs of his machine gun.

After a few intense moments, the shooting waned, then stopped. It was so strange, the way the Chinese fought. They were ethereal. As quickly as they appeared, they slipped into the shadows.

"Don't worry," yelled Yancey. "They'll be back."

Yancey had his men inspect the Chinese bodies. On one of them they found a tape measure, a plotting board, and a surveying tool called an alidade. "Probably a scout," Yancey said, figuring the man had come to diagram Marine positions for Chinese mortarmen downslope. Papers on his person identified him as an officer in the Seventieth Division.

At that moment, a sniper fired a round from long range. A spent bullet grazed Yancey's left cheek and drove deep into the soft tissue of his nose. Cursing, he hocked and spluttered as blood coursed down his face and into his mouth. The gash smarted terribly, but Yancey was okay. Refusing medical attention, he methodically took off a glove and snatched the sliver of metal from his snout. The blood instantly crusted over and froze to his bearded skin.

Yancey looked like a wild chieftain, smeared in war paint. He scanned the slope and fumed. He told his men to dig their holes a little deeper. "They'll be back," he said. "But we're ready for them, understand? Just do what I tell you."

19

BOON COMPANIONS

Toktong Pass

The company captain rocked on his haunches as he surveyed the hill, trying to foresee how he and his 245 men could possibly defend it. Manchurian winds keened through the mountain pass, and snow crystals glittered on the air. The captain's men had not arrived yet. They would be coming later in the afternoon, hauled up in a convoy of nine trucks from Hagaru. For now, he was alone in this wilderness, swallowed by it. He had been dropped off in a jeep an hour ago and had insisted on staying here by himself so he could have some time to think. He squatted on the brow of the hill, taking in the shadows as they shifted in the afternoon light. He had spent the past hour selecting this place, staking this ground. It was just a foothill, stippled with brush that scratched in the wind. Yet he knew instinctively that, should a battle take place around the Chosin Reservoir, this spot could be the most important swatch of terrain. That was why on this day—November 27—Captain William Earl Barber, commander of Company F (Fox) of the Second Battalion of the Seventh Regiment, had been sent from Hagaru to this isolated spot beside a hairpin crimp in the road. It was his job to map where his men might dig in for the night, to memorize the angles and sight lines, the advantages and vulnerabilities, the possible fields of fire.

Bill Barber was a wily chess player. He loved and understood strategy. A crack marksman, he had fought bravely and resourcefully at Iwo Jima, where he won a Purple Heart and a Silver Star. Barber was thirty years old now, a tough, dough-faced, God-fearing man who spoke in a twangy eastern Kentucky drawl. He had grown up in a little place called Dehart, Kentucky, near the Licking River, the son of subsistence farmers. This place reminded Barber of the hard-scrabble country around his home—the same hodgepodge of ridges and draws, the oddly amplified sounds that would carry down the hollows, the way the land rolled on and on. It possessed the random quality of a crumpled sheet of paper. The one feature that gave the country logic, that tied it together, was the road worming through it.

Toktong Pass was the highest spot along the main supply route. At this place, the narrow road cut along the shoulders of Toktong-san—a domed mountain that, at 5,454 feet, was the highest point anywhere around the Chosin Reservoir. This pass was a funnel through which Smith's mechanized forces would have to travel on their way north. Its strategic importance was inescapable. If the Chinese were out there somewhere, in large numbers, they could seize the pass and chop the First Marine Division in two. The regiments at Yudam-ni would then be cut off from the men down at Hagaru. The battle would practically be over before it had begun.

If the Chinese chose to take this eminence, one company was a woefully insufficient force to hold them off. Barber would have to apportion his assets wisely. He would have to create a perimeter that could not only protect the road but also protect itself—for an attack could come from any direction of the compass. Now, as he walked the hill, a layout began to take shape in his mind: where he would emplace his 81-millimeter mortars and his water-cooled heavy machine guns; where his three rifle platoons would dig in; where he would put his own command post, the medical tent, the warming tent; where the communications wire would have to be strung.

Most people back in Hagaru seemed to think that putting Fox Company out on this hill was probably an unnecessary precaution. The attitude, handed down from Almond and still permeating the ranks, despite Smith's own circumspection, was that the few Chinese in the area had come simply to monitor, stall, and harass the Americans. They weren't a serious fighting force.

But Captain Barber thought otherwise. He had taken the time to read a tract that had been found on a captured Chinese soldier and reworked into English by Army translators. *Military Lessons,* as the pamphlet was titled, belittled American fighting abilities. "Their infantry is weak," the tract declared. "These men are afraid to die, and will neither press home a bold attack nor defend to the death. If their source of supply is cut, their fighting suffers, and if you interdict their rear, they will withdraw."

Barber had also studied the military strategies of Sun Tzu, the legendary Chinese philosopher from the fourth century B.C. No one was a more diligent follower of Sun Tzu's precepts than Mao, whose military treatises Barber had also read. During the civil war, Mao had often pursued a strategy of camouflaging his own forces in a way that would embolden the enemy, enticing him to come forward. Then, at an opportune moment, Mao would encircle his foe and destroy him.

Barber wondered whether the Chinese at Chosin might be employing this same policy: hiding, luring, circling, pouncing. Straight out of Sun Tzu's *The Art of War.* Barber sensed that, even now, enemy soldiers were crawling over Toktong-san, watching him.

≡

With a mashing of gears, Barber's company appeared, grinding up the valley. The 245 men from Fox unloaded on the road and climbed the hill with their belongings and their tools. They brought with them quite an arsenal: M1 Garands, carbines, Browning Automatic Rifles, light machine guns, mortars, bazookas, .45 caliber sidearms, and countless bandoliers of ammunition. By the time the men had humped their stuff to the top, it was nearly five o'clock, and the light was failing. They didn't have much time to dig themselves in. They gathered on the hill—which everyone was simply calling Fox Hill—and waited for instructions from Captain Barber.

These Marines had only become acquainted with Barber over the past two weeks. In truth, they didn't much like him at first. Assigned to lead Fox Company after the Inchon landing and the attack on Seoul, Barber had come over fresh from Japan in mid-November. For their tastes, he had seemed too clean-cut, too spick-and-span. One Fox Company Marine thought Barber looked like a "candy ass."

Another thought he was "all dressed up like a well-kept grave." The first day, he had gathered the company for a little speech in which he boasted how much he knew about tactics. "Frankly," Barber had said, "I'm a hell of a good infantry officer."

The men of Fox Company didn't know anything about Barber's intrepid performance on Iwo Jima—how he'd come ashore as a twenty-five-year-old lieutenant and, within a few short weeks, had found himself leading not just a platoon but a rifle company, in some of the most horrendous fighting the Marines had ever faced. He was shot in the hand, then suffered a nearly catastrophic concussion, blood pouring from both ears. He was evacuated, but he recovered and returned to the battle, where he endangered his life crawling through enemy cross fire to rescue a couple of wounded men. Barber fought to the bitter end.

Later, he spent several months in Japan with the occupation forces. During peacetime years, he stayed with the Marine Corps and bounced around the United States, serving, among other things, as an instructor, a recruiter, and a rifle company commander. He was what the Marines call a "mustang." Mustangs are officers who began their careers as enlisted men and earned their commissions by working their way through the ranks and proving themselves in combat. They're officers who, as a result of their grunt pedigree, often seem to have a comprehensive perspective that gives them a special confidence in the field.

Certainly that seemed true of Barber. The day he met the men of his company, down near Hamhung, he had found much to be desired. He thought they looked raggedy and lackadaisical—like a bunch of Pancho Villa's banditos, as he put it. Barber's first order was to make his men shave. He made them clean their weapons, too, and then enforced a series of exercises in remedial marksmanship and basic physical conditioning. It was as though he were putting them through boot camp again—although, in truth, many of them had never been through boot camp the first time, or at least their original training had been improper and rushed.

The men resented this new martinet in their midst who treated them like bumbling novices. But the weaknesses he saw in them were real, a fact that many in Fox Company would grudgingly come to admit.

"Luck in combat is fickle," Barber once said. "But I've noticed through the years that those who make the best preparations enjoy the best luck." Making preparations was what Barber, always the stickler, remained focused on tonight. Now that the sun had dropped behind the mountains, the temperature had slipped well below zero. The men were stiff from their seven-mile, open-air truck journey up the valley, and they were looking forward to taking turns in the warming tent, sipping hot coffee and maybe enjoying some half-boiled C rations.

But Barber's first order scotched that possibility. He gathered his platoon leaders and, with some urgency, handed down the plan: The men were to form an oval-shaped perimeter along the hill and start digging foxholes—immediately. There would be no warming tent tonight, and no fires. Everyone would be sleeping on the ground, at alert, watching for the enemy. Many of the men cursed at this order— they'd seen no signs of the Chinese the whole way from Hagaru. This seemed yet another unnecessary precaution from their annoyingly scrupulous captain. People were likely to freeze to death on the exposed hill.

An order was an order, though, and the men followed their platoon leaders, trudging over to their assigned positions. The members of Fox Company broke out their entrenching tools and started hacking into the frozen earth. From his command post, which was starting to take shape near the road, Barber was pleased to gaze up through the blue gloom and hear the concerted sound of many metal blades scraping the iron-hard ground.

≡

Far up the hill, on the perimeter's northwestern edge, the men of Barber's Second Platoon were trying to chisel out their ring of foxholes near a scattering of granite boulders. Among the thirty members of the platoon were two friends from northern New Jersey—Hector Cafferata and Kenneth Benson. The two men had vaguely known each other back home, had competed against each other on the gridiron, but now they were close pals and foxhole buddies, halfway around the world. Cafferata and Benson spent an hour chipping at the ground, but their exertion produced only a pitiful divot a few inches deep. "Jesus Christ," Cafferata said. "You'd need a goddamn stick of dynamite to make any progress here."

Declaring the excavation project pointless, they decided to build a windbreak instead. They cut some brush and pine saplings and then piled the limbs in a semicircle around their sleeping bags. They anchored the makeshift screen with a few well-chosen rocks—and called it good enough. This was home for the night. They unrolled their mummy bags and cussed Captain Barber for making them freeze their asses off out here, when it probably wasn't necessary at all.

Private Hector Cafferata, twenty-one years old, was a strapping guy of six foot three. He had big hands, clompy feet, a bulbous nose, and the foulest mouth in the company. People called him "Moose" or "Big Hec." He grumbled a lot, making a rheumy, gravelly sound deep in his throat. He was goofy, stubborn, maybe a little boneheaded, and he often got himself into trouble. He had a reputation for being the biggest fuckup in Fox Company. People loved him just the same, because he was full of bumptious energy, and he always had a good story to tell.

No one could gainsay his marksmanship. Big Hec was an incredible shot. He had grown up with a gun in his hands. An outdoorsman from age twelve, he especially loved duck hunting, rising before dawn and heading to the wetlands. Gray weather, freezing water, dark skies overhead—it didn't seem to bother him. He'd shoot a duck and it would fall through the shell ice. He'd remove his clothes and wade out naked, busting through the frozen skim to retrieve his quarry, then wade back to shore, wipe off the freezing water, climb back into his clothes, and head for school—where, by agreement, the administrators had him store his gun, his birds, and his sometimes bloody jacket in a custodial closet. "The outdoors, that was home," Cafferata said. "And I could shoot. *Pfft. Pfft.* I knew how to put the bullet where it belongs."

When he was growing up, Cafferata had two heroes: the heavyweight fighter Joe Louis and the U.S. Marines. "I was one Marine-happy kid," he said. "The Marine thing, I wanted it. Whatever that is. The idea that you'd give your life for him, and he'd give his for you. It was a mystery to me. I was always gonna be a Marine."

Ken Benson, on the other hand, had never been much of a hunter, had never been particularly drawn to guns. Mainly, he was a sports nut, and quite a gifted athlete—basketball, baseball, football. You wouldn't know it by looking at him. He wore a pair of thick glasses that made him look like a nerd. Private Benson—Bense, as he was known—was only nineteen years old. He had grown up in Newton, New Jersey, in the Kittatinny Valley, about fifty miles from New York City. He was a funny guy, orderly in his thoughts, a font of sports trivia. Something about him calmed Cafferata down. He understood Big Hec, looked after him, knew how to let the excesses of Cafferata's sometimes outrageous personality roll off his back. The two New Jersey kids were always giving each other hell, joshing and cursing and shooting the breeze. But they were inseparable friends, boon companions.

They removed their boots and squirmed into their mummy bags. The temperature was twenty below zero. The mountain air tingled in their nostrils. Their barrier of pine saplings helped only a little to buffer the biting wind. Still, Fox Hill had a kind of haunting splendor in the bright moonlight, everything suffused in a cyan haze. Looming above the scene was Toktong-san, a stark citadel of granite. The two men could hear the brittle clinks and shuffling sounds of the company hunkering down for the night. As he dozed off to sleep, Hector Cafferata could feel his M1 resting on his chest.

20

EASY COMPANY HOLDS HERE

Yudam-ni

A few anxious hours passed, but all remained calm on Hill 1282. Maybe Yancey and his men had been spared? Maybe the Chinese, faced with the Marines' stout firepower, had had second thoughts? Maybe the weather was just too much? In olden days, when it got this cold, warriors laid down their arms and said, "See you in the spring." A gentleman's agreement.

The men of Easy Company relaxed. Some of them began to doze off. Then it started. Shortly after midnight, Yancey heard a mashing noise, a queer and disconcerting sound that was both delicate and huge. It sounded, he thought, like thousands of feet walking across a carpet of cornflakes. It took him a while to grasp what it was: The Chinese, hundreds of them, were stamping on snow, in a brisk cadence, moving into place beneath Hill 1282. Alarmed, Yancey cranked the handle of his field set and reached Ray Ball, the company executive, who was somewhere on the rearward slope, near Captain Phillips.

"They're coming up the hill, Ray," Yancey reported.

"You sure?"

"I can hear the fuckers crunching through the snow. How about some illumination?" Yancey wanted the mortar guys to lob some star shells overhead so he could see what he was dealing with. The moonlight, bright as it was, did only so much.

Ball said they were running low on illumination rounds, but he'd do what he could. By now Yancey could hear an even more bizarre sound. Some Chinese drill sergeant was down there, crying out in the night, and his chant spread among hundreds of approaching enemy soldiers. The words were uttered in heavily accented English:

> *Son of a bitch, Marines.*
> *We kill!*
> *Son of a bitch, Marines,*
> *You die!*
> *Nobody lives forever.*

The chant, uttered over and over, carried eerily on the wind. But it was drowned out by a louder sound: a ghoulish din of cymbals, drums, bugles, whistles, and bleating shepherds' horns. This was how the Chinese units, lacking radios, signaled to one another across distances. The strange bedlam sounded, said one account, like "a witches' conference." It was, said another, a "terrible moonlit serenade, a nightmare scene, a lunatic's delight."

Finally the requested star shells boomed in the sky and lit the hillside in the harsh blue glare of phosphorus. The scene was terrifying. It looked like a whole battalion of Chinese, swarming up the slope toward Yancey's platoon. Off to the left, another enemy contingent screamed down from the ridge and bounded straight for Gallagher's machine gun. "They came in a rush," said one Easy Company rifleman, "like a pack of mad dogs." Another said it was as though the snow had "come to life."

The men crouched a little deeper in their holes and looked to Yancey for cues. "Lieutenant Yancey," one of them said nervously. "How many Chinese're in a horde?"

Yancey directed the platoon to hold fire until the enemy drew closer. "All stand steady!" In their foxholes, the men began to take aim. Their adrenaline was surging, and their fingers fidgeted on the triggers. The Chinese were almost within range. Along the line, the trip flares started popping off, illuminating individual pockets of attackers. The chanting droned on:

> *Nobody lives forever*
> *Marines, you die!*

By this point, Yancey had had enough. "That's right," he snarled back. "No one lives forever, you bastards!" He leveled his carbine and fired a full clip. Then the whole platoon opened up. From the foxholes came a deafening fusillade of BAR and M1 fire. Tracers zinged over the slope, throwing off a red gleam. Then it was the machine gunners' turn. "Let 'em have it!" Gallagher screamed. He mashed the trigger on his .30 caliber, and the others did the same. The barrels of their weapons glowed orange from the heat of constant firing.

The Chinese fell in long rows, but new waves of soldiers were rushing right behind. They blasted away with their burp guns and Thompsons, kicking up the snow in stutters. When one phalanx was cut down, the next would crawl over the bodies, sometimes grabbing a dead comrade's weapon. Yancey's Marines couldn't comprehend them: Either they were inordinately brave, inordinately stupid, or inordinately fearful of their own superiors, for they kept advancing, with no apparent regard for their staggering casualties. "There was just so many of 'em you could kill," said Easy Company private Robert Arias, who was part of a machine-gun crew. "If you had a conscience it was hard, because you knew they had a family back home. But you had to kill them—sometimes at point-blank range. Just as soon as you mowed them down, they'd come back in another wave."

So wild-eyed and manic were their sorties that many a Marine came to believe that the Chinese must have been hopped up on some powerful stimulant. As they crept closer, the Chinese began to hurl concussion grenades at Yancey's platoon. The sharp smell of picric, the acid used in their crude explosives, wafted down the line. The platoon was taking a beating now. In several places, the Chinese were starting to penetrate the perimeter. Yancey raced toward a weak spot to encourage his men. "Keep firing!" he urged them. "Don't let 'em through!"

When the action lagged for a moment, Yancey looked around and realized that the battle wasn't raging only here on Hill 1282; it had broken out across North Ridge and beyond. The highlands around Yudam-ni had come alive with pyrotechnics. It looked as if the mountains were covered in fireflies.

Yancey picked up the field phone and reached Ray Ball again, at Phillips's command post. "Lay it on, Ray!" he said. "Lay it on!" Ball understood the request and had the mortar guys go to work, lobbing 60-millimeter shells up and over Yancey's position. The message was relayed down to the valley, too, and soon the howitzers were pounding preregistered targets along the ridgelines beyond Hill 1282.

When the shells hit, plumes of snow and clods of debris were ejected from the slopes. Yancey could hear a chorus of wails and moans as Chinese bodies were hurled skyward. Again and again, the incoming rounds whined overhead, and shrapnel sawed through the icy air.

The next star shell revealed a horrific panorama: The snow was smeared with blood. Twisted corpses and shorn body parts had been flung in all directions.

≡

From his command post, Captain Walt Phillips, the commander of Easy Company, was following the action on 1282 with increasing alarm. All three of his platoons were in serious trouble; the fighting along the ridgeline was more ferocious than anything he'd ever seen. The Marines were throwing everything they had at the enemy, but it wasn't enough: The press of the Chinese numbers, the sheer weight of their onfall, was overwhelming. The CCF wanted that hill. Their commanders had decided it was of the utmost importance to seize it. Now the survival of an entire company of Marines hung in the balance.

Phillips saw that it was the Second Platoon—Yancey's—that was taking the brunt of the attack. Gun emplacements were being overrun, the dead were piling up, and medical corpsmen, who'd found they had to put ampoules of morphine in their mouths to keep them from freezing, were treating the wounded by the score.

At about 1:00 a.m., Phillips broke away from his command post and trudged the forty yards up the steep hill to help bolster Yancey's position.

"Where's Yancey, where's Yancey?" he demanded as he roved Second Platoon's line. Phillips finally located Lieutenant Yancey, covered in fresh blood: A grenade had exploded not far from him and blown him twenty feet across the hill. A fragment had lodged in the roof of Yancey's mouth, and another one had sliced into the bridge of his nose—his second facial injury of the night, this one far more serious than the first. Yancey could hardly talk, and he struggled for air as blood trickled down the back of this throat. He kept gasping and snorting to clear his clogged airway, and he looked to be in excruciating pain. But he was still fighting alongside his men, grunting orders that were barely decipherable, firing his carbine from his hip.

Phillips joined Yancey along the perimeter. "Doing well, men," the captain tried to reassure them. "Stay loose, Marines, you're doing fine." Phillips soon suffered his own injuries: He was shot in the shoulder, then in the leg, but he kept with the men.

The platoon was faltering. More than half the ranks were either dead or wounded. Yancey, brandishing a .45 revolver now, kept trying to rally the platoon. Repeatedly, the Chinese burst through the lines, only to be driven back in spasms of close-in fighting. The battlefield was all clamor and confusion, with random scrums of men locked in mortal combat. Yancey's boys were using their bayonets and firing point-blank. They were fighting with fists and pistols, knives and entrenching tools. But the Chinese kept boring in.

Staff Sergeant Robert Kennemore, a machine gun section leader, was scrambling across the hill. "Don't go down there, you fool!" Phillips yelled after him. But Kennemore ignored the captain. He thought he had seen a group of Chinese dragging away one of his machine gunners by the legs, bludgeoning him, stabbing him with bayonets. Recognizing that he was too late to help, Kennemore crawled toward a gun position on the far right. He started searching through the bodies of the Marine dead, hunting for anything useful, when a Chinese grenade plopped beside him. He seized it and flung it just in time. But another one landed nearby, in a gun pit occupied by three Marines. Then a third grenade dropped beside the second one.

Acting on an impulse of pure selflessness, Kennemore stepped on one grenade and squashed it into the snow, while at the same time crouching on a knee to absorb the blow of the other grenade. How, in those fleeting moments, he saw the angles of the situation, no one could understand. The two grenades exploded almost simultaneously, and tore into him. The three crewmen in the gun pit, though temporarily deafened by the blast, were spared.

All of North Ridge was in peril. Captain Phillips was sure the dike was about to break. He knew he had to do something. As corpsmen tended to his wounds, he got on the phone and called for reinforcements from battalion headquarters in Yudam-ni. Phillips asked if there was enough time to scramble a few platoons and march them up the slippery hill to save the company, but he couldn't get a clear answer. Easy would have to stay firm and hope for a miracle. But the captain was determined not to budge from the crest.

As if to make his point, he took a bayoneted rifle and planted it into the hard soil of 1282. "This is Easy Company!" Phillips cried out. "Easy holds here!"

A few moments later, a stitch of enemy bullets rang out, and Captain Phillips dropped dead. Beside him, the rifle creaked as it swayed in the wind.

21

WHERE THE BULLET BELONGS

Toktong Pass

Hey, Bense! The fuck is that?" At around 1:00 a.m., Cafferata thought he heard the crackle of gunfire down the hill, maybe on the road. "Jesus Christ. Somethin's happening." Benson tried to pull himself awake.

Then Cafferata heard something closer, a crunching sound on the snow. He popped out of his bag, standing in his stocking feet, and spied six Chinese soldiers walking up the hill right in front of him. He could see them in the moonlight and by the eerie glow of the flares. He leveled his M1 on the first one and aimed for his chest. He fired; the soldier dropped. Then Cafferata shot the five others. Benson, through all this, was still struggling to get his boots on. "What the hell you doing, Bense? Fuck the boots. You better start shooting!"

Benson got to his BAR and started firing at the moving shapes in the distance. Cafferata could see more of the enemy marching uphill. "We can't hold 'em," Cafferata said. "We gotta fall back. We don't have no time to fiddle."

Cafferata and Benson were more than two hundred yards from the company's main line of resistance. They got on all fours and started to crawl through the snow. As they were scrabbling along, an object hit Cafferata squarely in the back. It rolled down his jacket, down his right leg. He turned around and saw Benson staring, saucer-eyed, at a Chinese grenade. It was a small explosive, encased in bamboo, with a glowing fuse made of cloth.

"Throw it, Bense! Throw it!"

Benson grabbed the offending missile, but when he hurled it, it caught on the lip of a little berm in front of them. The grenade detonated and sent up a cloud of ice shards, frozen dirt, and splinters of bamboo. Benson's ruined glasses were thrown into the air as the ejected debris lodged in his eyes. His face was burned, cut, and bloodied.

"God damn, Hec. I can't see a fucking thing."

Cafferata looked around to learn where the grenade had come from. He could see scores of Chinese soldiers, shadowy forms moving over the snow. "We got to get outta here!" Cafferata whispered. "Hold on t'me, Bense." They crawled along for ten yards or so, with the now blinded Benson clutching one of Cafferata's big feet. They reached the next set of foxholes and realized that the three Marines inside them weren't moving. The Chinese had come along and shot them in their bags.

The two men continued until they reached a little draw, where some wounded Marines had taken cover. Cafferata decided they would stay here—this was where they would make their stand. The enemy kept coming, only now they were tooting bugles, crashing cymbals, blowing whistles. Cafferata smashed two of them with his entrenching shovel. He scooped up a Thompson submachine gun one of them had dropped and sputtered at the next line of attackers.

Then he grabbed his M1 and began to shoot in earnest. While Cafferata fired away, Benson sat in the snow beside an ammunition box and fumblingly loaded spare M1s and carbines taken from the dead and wounded Marines in the draw. Cafferata found that the carbines didn't perform well at all; their firing mechanisms couldn't stand the cold. But the M1, the widely revered semiautomatic rifle with a clip housing eight .30 caliber rounds, proved reliably lethal.

The two Jersey boys worked as a team, almost as a single organism. Benson got faster at loading, more efficient. He summoned the muscle memory in his fingers; he didn't need to see what he was doing. When Cafferata spent his eight rounds, Benson would be ready with the next freshly loaded M1.

Though Benson couldn't see it, the scene around them was surreal, spectral. Tracers snapped through the air, flares arced across the night sky. It seemed that an entire Chinese regiment had descended on them. All of Fox Hill was lit up, but Cafferata could only focus on his piece of terrain. He kept shooting, and the Chinese kept dropping. At one point he had to leave Benson to repulse an attack on the other side of the draw. His M1 grew hotter, accumulating carbon, sometimes blurting off two shots instead of one. Smoke curled from the weapon, until finally the barrel guard caught on fire. He smothered it with snow, and it ticked and steamed as it cooled.

In back of his fury, Cafferata felt sorry for the Chinese. He couldn't understand why they kept running headlong to their deaths, as though they wanted to get it over with as soon as possible. Some of them charged at him with quaintly crude weapons, almost archaic in some cases. One enemy soldier had a long pole at the end of which a knife had been attached with string. In other cases, they charged with no weapons at all. He wondered where that kind of bravery and fanaticism came from. Did they do it for love of country? To defend an ideology they held dear? To assert some principle held deeper within their society? Or did they do it because their officers compelled them to?

He couldn't understand why their rounds didn't find him—him especially, a big, lumbering form silhouetted in the moonlight. Over and over again, he had made a target of himself. He could see their muzzle flashes, could see their blood spattered on the snow. He could hear their moans on the slope below him. He could feel their bullets slicing by his head, winging around, ricocheting. Yet it was mystifying—he wasn't touched.

Benson, still loading weapons on the ground, thought that his blindness had accentuated his other senses. For him, the battlefield swirled with queer sounds and pungent smells. He listened intently to the rising and falling of the Chinese war cries, their pealing bugles and shepherds' horns, the drums pounding in the distance. The whole hill seemed to quake. On the shifting winds, Benson could smell the sour stink of phosphorus and cordite. He swore he caught the reek of garlic permeating the clothing on the bodies of the enemy dead. Even their gunpowder had a strange odor, he thought—some said the Chinese lubricated their weapons with whale oil.

The grenades started coming in clusters, and Cafferata surmised that a bunch of Chinese soldiers were downslope, just out of view behind some rocks. They had to be lying on their backs, he figured, and hurling the grenades backwards up the hill. As the projectiles sailed in, Cafferata picked up his entrenching shovel with both hands and swatted them back upon their throwers. He had never been much of a hitter in baseball, but now, manic with adrenaline, he swung with surprising accuracy, whacking away the grenades, one after another.

≡

With Benson at his side, Cafferata kept fighting through the pre-dawn hours. He had never killed a man until this night; now he had killed dozens. Shooting the Chinese at a distance wasn't a problem for him—it didn't seem much different from duck hunting. But when he could make out their faces and see the fear in their eyes, that was another thing. They were kids, like him. They didn't want to be here any more than he did. But what could he do? He had to shoot them.

The bodies piled around him—he was using them as a screen. It seemed ghastly, in a way, protecting himself from the enemy with the enemy's own corpses, but it was the only cover he had. (Other Marines throughout the Chosin battlefield would do the same thing—they came to call the corpses "chop suey sandbags.") The Chinese bullets would thud into the stiff flesh. A grenade would land in their midst, and frozen bits of sinew and bone would fly through the air, scattering over Cafferata and the wounded Marines.

During a lull in the fighting, Cafferata began to think what a waste it was. He had no reason to hate these people. It occurred to him that if they could pool the money and the resources it had taken both nations to assemble all these kids out on this hill, they could have signed on for a terrific pleasure cruise instead, Americans and Chinese alike, somewhere in warm tropical waters, and they would have had a ball.

Then another grenade hissed overhead and landed beside Benson. Cafferata stooped to pick it up, but it detonated as it left his grip. The blast mangled some of the fingers of Cafferata's right hand, peeling tissue off the bone. In the moonlight, he thought the exposed flesh looked like meat straight from the freezer. Enraged, he cursed and yowled and spat. His fingers were shredded. He'd have to pull the trigger with his thumb.

Cafferata resumed his position and took aim. Benson was beside him, attentive to the noises around him, his fingers growing ever more dexterous. Though he still couldn't see a thing through the blood and shards and grit in his eyes, he remained poised, and this had a soothing effect on Cafferata. Through the fighting, Cafferata perceived that everything was happening in rapid succession, almost too fast to process—and yet it all seemed slowed down, dreamlike. A wave of panic would wash over him, and he would summon every nerve to conquer it. He was terrified, but at the same time, he was functioning automatically, without thought. Every bird he'd ever killed, every deer, every varmint in the pine barrens, in the marshes, every day he'd spent shooting at some stupid target—all of it was embedded in his experience, in his reflexes. *Aim straight, and put the bullet where it belongs*—that was what mattered now.

He felt, surging within him, a frantic debate between his feet and his consciousness. His feet kept telling him to run, but his mind kept telling him there was nowhere to go, that he had to stay here, that he had to keep shooting or else he and Benson and the others would surely die. "There's no place to hide," he said. "You've got a choice: Kill or be killed." He was hyperalert, consumed in a battle frenzy. True to his name—Hector—he had become a warrior. "I was hopped up," he said. "The adrenaline was flowing." He dropped thirty, forty, maybe fifty Chinese soldiers, yet he remained unscathed by their bullets. The first tinting of dawn seeped into the night sky, and still they kept coming.

22

GUNG-HO, YOU COWARDLY BASTARDS

Yudam-ni

Through the early-morning hours, Private Stan Robinson lounged comfortably on a stretcher in a cozy medical tent lit by Coleman lanterns and bathed in the steady heat of a kerosene stove. But he was feeling restless, useless, and blue. *Frostbite?* It seemed such a pathetic excuse. Robinson knew he should be in the hills with Yancey and the platoon. He could hear the distant booms of battle, and he wondered how Easy was faring.

Then an ambulance jeep pulled up and a bunch of bloodied men, fresh from the ridge, were hauled into the infirmary. "What unit you in?" Robinson asked one of them.

"Easy Seventh."

Robinson's ears perked up. "We get hit?"

"Creamed," the injured man said. The situation was bad up there, and Easy was in serious trouble. "Yancey's wounded," he added. "Everybody is, I guess."

That spurred Robinson to action. He climbed into his soiled clothes and field parka and gingerly slipped some boots onto his hideous, swollen feet. He parted the tent flaps and hobbled out into the cold night. He grabbed the first spare rifle and cartridge belt he could find and was exiting the compound when a medical corpsman accosted him.

"Hey, Robinson—back inside!"

Robinson glowered. "Get the fuck outta my way." He limped from Yudam-ni and started toward North Ridge. It took him an hour or more. He slipped and crawled up the slope like an old wino, crying out in pain when his blisters burst. It felt as though he were going against the current: On the way up, he kept encountering litter bearers, hauling serious battle casualties, who were headed for the same medical tent from which he'd just absconded.

But finally Robinson reached 1282. This was where he belonged. The battlefield, his platoon mates liked to say, was Robinson's one true home. He found Yancey standing beside a machine gunner. The Chinese, at the plaintive blare of a bugle, had pulled back into the night, but everyone knew they would return.

Robinson, with a shit-eating grin, tapped Yancey on the boot.

"Well, I'll be damned," Yancey said, admiringly. He spat a gobbet of dried blood onto the snow.

Robinson said he was looking for a job.

"Over there," Yancey said, and put Robbie to work.

≡

The next assault began around 3:00 a.m. The Chinese must have sensed that Yancey's platoon—indeed, all of Easy Company—was at the breaking point. They attacked with ferocity. One wave struck, then another, then another still. Private James Gallagher hammered away with his machine gun, and Robinson proved lethal with his BAR. Yancey made a perfect target, stomping around in the garish light of the flares, barking orders. The corpsmen kept trying to stop him and treat his wounds, but he wouldn't rest long enough to submit to their ministrations.

On the hillside, the Chinese bodies were hurled like matchsticks. Hundreds of them were sprawled in the snow, frozen solid in weird contortions. But the Chinese replacements kept stepping forward, as though churned from an assembly line.

Yancey knew he was running out of men. He kept tightening his perimeter, concentrating his fire. He had lost more than fifty dead and wounded, and he had no idea if help was on the way. He was incommunicado: The Chinese had severed his phone lines, and his radio had been smashed beyond repair.

Alarmed by a new breach in the line, he pulled together a group of nine guys for a counterattack, barking, "Marines—follow me!" Robinson was at his side, but no one else followed. They were too scared or too addled by shelling to budge. Yancey flew into a rage. "Gung-ho, you cowardly bastards! I said 'Follow me!'" Finally he got them moving, and they plugged the hole in the line.

In the midst of this action, a Chinese attacker, armed with a Thompson submachine gun, moved in close and squeezed a burst into Yancey's face. One of the bullets entered Yancey's cheek just below his eye and angled down through his sinus cavities, coming to rest in the back of his neck, near the base of his skull. Luckily, it had missed his spine, but along the way the projectile had fractured his cheekbone, jarred loose numerous teeth, and dislocated his jaw. Worst of all, the bullet pried Yancey's right eye from its socket. The ball, dangling by a cord of nerves and fibers, rested high on his cheekbone.

Yet, somehow, Yancey was still functioning. He picked himself up from the snow and, with his good eye, spied his assailant reloading his Thompson. Reflexively, Yancey snatched his .45 pistol from its holster (a weapon he had taken off a dead Japanese officer eight years before at Guadalcanal) and fired two rounds into the enemy soldier's abdomen, killing him. Then Yancey, horrified but not knowing what else to do, cradled his eyeball and gently mashed it back into its ragged hole.

Yancey was preparing to resume the fight when along came a godsend: A relief force from Charlie Company, Fifth Marines, led by Captain Jack Jones, had marched up from Yudam-ni to drive the Chinese off the hill. It would take hours of combat before North Ridge would be fully retaken, but Yancey's trial was over. He relinquished command of the platoon, turned over Hill 1282 to Jones—and collapsed. His unit had suffered 90 percent casualties.

With the appearance of the sun, a Corsair from the First Marine Air Wing, based on the coast, came roaring overhead. The pilot flapped his wings, then joined the battle, scattering the Chinese attackers below. As this first morning would show, Marine aviators, as well as Navy pilots flying fighter planes from carriers, would play a significant—and deadly—role in the daylight hours of the Chosin

Reservoir battle, often dropping preposterously low in "close support" of Marine actions on the ground.

The last survivors of Yancey's platoon went onto the field to identify the dead. A cry came from the hill: "Corpsman! Corpsman!" Within the tangle of bodies, a Marine was moving: It was Sergeant Robert Kennemore. His platoon mates had assumed there was no way he could have survived the blast of two simultaneous grenade explosions—but he had. At some point during the night, he had awakened. He'd felt for his legs and found they were useless, a shred of sinew and exposed bone. He'd stabbed himself with a syrette of morphine and then summoned the strength to drag himself a hundred painful yards across the hill, to a place where he would be more easily spotted. No one could believe it. "Christ, it's Kennemore!" They put him on a stretcher and carried him to Yudam-ni.

But Lieutenant Yancey, weak and ashen-faced from loss of blood, insisted on walking. His right eye wandered in its socket, and his broken jaw hung loose and floppy, like an unhinged gate. His teeth were shot into fragments that rattled around in his mouth. His parka was sieved with bullet holes and frayed from flying shrapnel. A topographical map in his coat pocket was later found to be perforated with burp gun fire. Yancey tore off a strip of blanket and wrapped it around his face, cinching the jaw into place. When he'd heard for certain that the hill had held, he gave his assent. A kind sergeant in front of him proffered a long stick and guided Yancey down the mountain.

The war was over for John Yancey. From Yudam-ni, he would be evacuated to Hamhung, and then Japan.

23

WHEN THE LEAD IS FLYING

Toktong Pass

At dawn, the fighting on Fox Hill began to subside, as it had at Yudam-ni. Hector Cafferata and the blinded Ken Benson had managed to hold on for nearly six hours, plugging what appeared to be a gaping hole between the Second and Third Platoons. The slope seethed with fog and smoke and the acrid smells of assorted propellants and explosives. The company had held, but just barely. Cafferata was willing to admit that Barber had been right: If the captain hadn't taken the precautions he did, if he hadn't carefully arranged the men out on the hill, Fox Company would have been wiped out. As it was, twenty-four Marines had been killed, more than fifty had been wounded, and three others were missing. Nearly a third of the company had become casualties in a single night.

The Chinese casualties, on the other hand, were more difficult to ascertain, but they were impressive. Along the perimeter, Cafferata could see snarls of bodies. Nowhere was the carnage heavier than right in front of his own position. There must have been a hundred corpses, he guessed. The worst of the attack appeared to have transpired where he and Benson had made their stand. The snow was dribbled and splotched with blood. In places, it looked like a Pollock painting.

Cafferata's squad leader, assessing the damage along the perimeter, came trudging by. He stood in awe of what Cafferata and Benson had done—he estimated that two enemy platoons had been destroyed. He had fought in the Pacific during World War II and had seen his share of gore.

"Sarge," Cafferata said. "Was it this bad on Okinawa?"

"Doesn't matter where you are," the sergeant replied. "When the lead is flying, *that's* the worst place you've ever been."

With a lull in the fighting, Cafferata was finally able to steal a good glance at Benson. The whole night, he hadn't had a spare moment to consider his friend or to apply any first aid. What an alarming sight Benson was. He looked like Oedipus—his eyes were sealed shut with scabby globules of grit, ice, and dried blood. His face was lacerated, bruised, and speckled, as though he'd crashed through a plate-glass window. Needles of bamboo were embedded in his skin.

"Jesus Christ, Bense," Cafferata said. "You look like shit."

Benson knew he had to get to the first aid tent. With the morning light seeping through gaps in his crusted eyelids, he thought he could see well enough to crawl toward headquarters. Cafferata agreed that Benson should get some help. In parting, he realized that, in addition to being his best buddy in Korea, Benson was the most profound friend he probably would ever have. You didn't go through an experience like that without feeling a connection. Cafferata's life was indelibly changed, and Benson was indelibly part of it.

Cafferata helped Benson on his way—he was worried about snipers—but he quickly returned to his post. The battle may have been ebbing, but it wasn't over yet. Across the slope, Cafferata could hear gunfire. In a few pockets, stubborn groups of Red soldiers continued attacking, charging at the Marines in the morning pallor. For now, Cafferata had to stay put, crouched in the same wash where he'd fought through the night.

=

Around 6:30 a.m., three Chinese soldiers approached Cafferata with their arms in the air. They were shaking, either from fear or from the cold, or both. Cafferata thought they looked like street urchins. They were short and scrawny, and couldn't have been more than sixteen years old. They had adolescent faces; some of their features seemed almost feminine. They cringed on the slope, smiling awkwardly, uttering something in Chinese, in obsequious tones.

Cafferata peeked from behind the lip of the wash. He waved his rifle at the three young men, signaling them to halt where they were so he could appraise them. They wore white quilted uniforms, floppy hats, and flimsy tennis shoes that appeared to be made of canvas, with soles of crepe. They were shoes that might make sense in a tropical campaign, but not here. The three Red soldiers kept darting glances at the sky. Maybe they were praying, but more likely they were watching for planes. The Chinese respected American airpower and feared that dawn could bring the prospect of terror from above.

These kids seemed pitiful. Cafferata recognized that, to them, he must have looked like a giant. Probably they were half-starved. Certainly they were freezing.

Then again, he thought, this could be a trick. Maybe their brothers-in-arms were hiding in the brush downslope, waiting to ambush him. He agonized over what to do. This was not a decision he was equipped to handle. He was just a private, twenty-one years old—low as they come, he liked to say, "lower than whale shit." After the night's fighting, he didn't trust his nerves. He feared that if he came out in the open to search the three Chinese, he might do something stupid, maybe get himself killed. He didn't know what protocol the Marines were supposed to follow in this situation. Captain Barber had never talked about what to do with prisoners.

Cafferata looked around at the bodies littering the slope. If he watched carefully, he could see that some of them were still writhing and twitching. Maybe they were in their death throes, or maybe they were just doing a so-so job of playing possum. He began to suspect that some of the Chinese were feigning death in order to lure the Americans near enough that they could open fire on them. A few of them did seem pretty good at impersonating cadavers: Sometimes the only telltale sign was a faint puff of vapor rhythmically rising from someone's mouth. Periodically, Cafferata could see an unarmed corpse spring to life and make for the brush. The nearest Marine would fire a few rounds. Some of the fleeing soldiers were hit; the lucky ones reached the safety of Chinese lines.

Cafferata considered the three prisoners again. Should he shoot them? Should he turn them loose? He didn't know.

He had to wonder: Why had these three men surrendered in the first place? They didn't seem injured. They hadn't been cornered or caught. They'd appeared out of nowhere and voluntarily turned themselves in to a lone American. It was as though, in a single night, they'd gotten their fill of fighting. Maybe they thought that if they laid down their arms, they might find some hot food.

Cafferata beckoned the three young soldiers to come forward. He gave them a cursory frisking and motioned for them to get on the snow beside him. He had them lie on top of one another, figuring that would limit their ability to make a sudden move, and that the heat of their commingled bodies would keep them warm. He gestured with his rifle and barked: "You screw around and I'll shoot."

$$\equiv$$

Captain Bill Barber had spent the night fighting for his life and the life of his company. His apprehensions had come horribly true. The Chinese had understood the strategic importance of this place, just as he had. They had struck hard from above, from the ridges and saddles, and they had struck hard from below, from the road itself. Barber was not one to say "I told you so," and he did not gloat over his prescience, but he recognized that if he hadn't ordered his men to dig in, everyone in the company would have been captured or killed.

In fact, Barber had nearly been overwhelmed in the first moments of the fighting. He had originally established his command post near the road, but the Chinese had hit with such a concentration of fire-power that he and his staff were almost immediately forced to abandon the headquarters and establish a new one, far up the slope.

Since then, Barber had been constantly on the move, patrolling the lines, shouting commands, firing his carbine, rallying his men. He hadn't made it over to where Cafferata and Benson had fought their fight, but it seemed he'd been everywhere else. He appeared to be indifferent to the dangers around him. Two of his runners had been wounded trying to keep up with him. At one point during a particularly blistering firefight, a sergeant had suggested to Barber that it might be prudent for him to take cover. Barber ignored him, saying, "They haven't made the bullet yet that can kill me."

"Barber was cool as they come that night," said Richard Bonelli, a private who served with Barber on Fox Hill. "He was quiet, no nonsense, firm. Thank God he was there. Without him, the Chinese would have run us over. Them sons of bitches—there was no end to them. They just wore us down."

Barber was impressed by the bravery and resolve of the Chinese soldiers, but he didn't think much of their tactics. He noted how they seemed to charge over and over again in the same place. He noted also how their battle cries forfeited the element of surprise: All those bugles and whistles and horns and cymbals sounded eerie, but by announcing themselves like that, they gave the Marines time to brace for every attack.

What shocked and even saddened Barber was how little the Chinese superiors appeared to value the lives of their own men. Many of them were sent into their charges unarmed—presumably they were supposed to scoop up the weapons of their fallen comrades. On Iwo Jima, Barber had seen plenty of crazy banzai charges, but an aspect of strategy usually revealed itself, a method in the madness. Here, he thought, the Chinese officers were just throwing souls away. From the initial counts, at least 450 of the enemy had died in the night's fighting on Fox Hill. Across the larger Chosin battlefield, the casualty totals for General Song's troops would be even more astonishing: His Ninth Army Group had lost ten thousand men during the first night of battle, a rate of more than one thousand casualties an hour.

But the Chinese had succeeded in at least one of their objectives: They had largely seized control of the road, and now they were erecting substantial barriers on both sides of Toktong Pass. They hauled down tree trunks, they rolled boulders into place, they blew up sections of the road. Barber understood that he was truly cut off. Except by way of air support, it was doubtful he could get any relief from either Hagaru or Yudam-ni. His men would have to hold here, indefinitely.

≡

Through the early morning, Hector Cafferata stayed with his three captives in the wash, scanning the slope for an attack. He spoke to his charges several times, making halting conversation, trying to soothe them—it was obvious they were scared to death. Then, at around seven thirty, a Marine from Captain Barber's command post came along. Word had reached Barber of the incredible firefight that had transpired here in the gap between the Second and Third Platoons. Stories had already begun to spread about Big Hec and his heroics: that he'd taken out a legion of Chinese, that he'd saved the lives of Benson and a half-dozen other trapped and wounded Marines, that he'd batted away grenades like Ted Williams. Cafferata, of all people—no one saw it coming. The biggest fuckup in Fox Company was already becoming a legend.

Cafferata ignored such talk. A hero, he'd once heard, was a person caught in the right place at the wrong time. It was more a matter of luck than anything else. He and Benson did what they did because they had no choice.

The Marine from headquarters asked Cafferata if he needed any help. "Yeah, take these guys," Cafferata said, pointing to the prisoners quivering on the ground. He felt a proprietary responsibility for them—they were *his* prisoners. He felt sorry for them and wished them well. The Marine prodded the three captives with his gun, ordered them to stand, and pointed them across the hill, toward Barber's tent, where, presumably, they would be interrogated. The Marine turned for a moment and said with a smirk, "Hey, Moose, what's with the socks?"

Cafferata looked down and saw, with some consternation, that he wasn't wearing his boots. They were still in his sleeping bag, where he'd left them the moment the shooting had started, more than six hours earlier. Without his knowing it, he'd spent the night fighting in his stocking feet.

He started to head over to his original position, by the windbreak he and Benson had constructed. Now that he knew his boots were missing, he decided he desperately wanted them—his feet were blocks of ice. But the way down to the windbreak was a gauntlet of horrors. He had to step through piles of entwined corpses. The Chinese casualties looked like wax dummies, their eyes frozen in terror. At one point, he passed by a soldier who seemed to be alive—Cafferata thought he saw his hand move. He knelt and, with a thumb, peeled back one of the man's eyelids. But the eyeball didn't flutter, and the pupil was rolled far up into his skull. The guy was dead after all; Cafferata figured he must have been imagining things. With a shrug, he shuffled on toward his sleeping bag.

He was astonished by the international miscellany of weapons to be found on the Chinese dead: There were Enfields from Britain. Shpagin burp guns from Russia. Arisaka bolt-action rifles from Japan. A few German Mausers. There were antique implements whose provenance Cafferata, who was obsessive about guns, could not guess. The most common weapons of all were Thompson submachine guns, the "Chicago typewriters" made infamous by Al Capone's gangsters. The United States had supplied Chiang Kai-shek's Nationalist forces with Thompsons for decades. After Mao prevailed in the Chinese Civil War, his Red armies had appropriated much of Chiang's arsenal—and, in many cases, had commandeered Nationalist troops themselves, put them in PLA uniforms, and sent them off to die in Korea. The irony wasn't lost on Cafferata: American weapons, carried by once pro-American troops, were being used to kill Americans.

Cafferata continued down the slope when he was startled by the sound of a grenade clunking on the frozen ground nearby. He hurled himself into the snow, trying to take cover, only to find to his relief that the grenade was a dud. Then he realized that it had come from the vicinity of the supposedly dead man he'd just examined.

"You rotten bastard!" Cafferata shouted. Something about what this enemy soldier had done, the brazenness and duplicity of it, had activated in Cafferata the most intense rage he had felt the entire night. Bic Hec was storming mad. He returned to the offending soldier and nicked his cheek with the tip of his bayonet. Curiously, this produced no reaction, so he fired a round into his shoulder. This jerked the man out of a most convincing performance. He bolted upright, to a sitting position, grimacing in pain. His eyes, now wide open, smoldered with spite and defiance. Some part of Cafferata had to admire him—he was proud and he was tough. This was one soldier who would never surrender.

Cafferata shot him twice in the head.

=

At Fox Company's command post, Captain Barber found some warm batteries that got the radio working, and he was finally able to reach his regimental commander, Homer Litzenberg, up at Yudam-ni. Litzenberg explained the larger situation as best he knew it: The Chinese were everywhere—in the hills, on the ice, along the road. Maybe more than 100,000 of them. Yudam-ni was surrounded. So was Hagaru, and Koto-ri, too. The entire First Marine Division was in peril. Litzenberg said he couldn't get Barber any reinforcements. He wondered if Barber could quit Fox Hill and make a dash north to the relative safety of Yudam-ni. It would be a difficult and dangerous march of seven miles. What did he think?

Barber said he didn't believe he could make it. He was worried about his wounded. The Marines had an ethos about this—you never, ever left them behind, no matter what. Barber didn't explain the problem to Litzenberg, not exactly. He feared the Chinese might be listening in on this radio frequency and might find his casualty report encouraging. But Barber did obliquely suggest that various "tactical necessities" dictated that he stay put.

"Colonel," he said, "I can't move, but I think I can hold the hill for another night. Any reinforcements you can get me, I'd be much obliged." Litzenberg said he'd do what he could but explained that things looked bad in all directions.

During the night's fighting, Fox Company had taken some prisoners, and through an interpreter Barber had begun to learn something about his foe: These soldiers were from a regiment of the CCF's Fifty-ninth Division. Many of them hailed from Shanghai and its surrounding provinces. They had been training for an amphibious attack on Taiwan when the orders were abruptly changed and they found themselves traveling north on a train bound for Manchuria. They were horribly ill-equipped for winter weather. Many of them lacked overcoats, and almost none had gloves. During the frozen nights of November, they had marched down through the North Korean countryside, each soldier keeping warm during rests only by pairing off with another soldier; these "hugging buddies," as they were called, slept entwined in each other's arms. The prisoners seemed willing to divulge hard information and were not chary with the details. Some of them volunteered that they were pro-American—that they loved the United States, in fact. Several of them had been Nationalists, they said, fighting for Chiang Kai-shek during the civil war.

Barber started to worry about the Chinese prisoners. By morning he had accumulated more than two dozen, and he didn't know what to do with them. He didn't have enough food to feed them. He didn't have wire with which to detain them, or extra manpower with which to guard them. There wasn't enough room in the warming tent, but if he left them outside to stand in the cold, they would freeze. On the other hand, if he sent them back to the Chinese lines, they would only return later that night with guns in their hands.

This was one of several major problems Barber had to stew over. Another one was ammunition: He was running out of it. Some of the Marines in the Second Platoon were down to a few rounds per man. He ordered work details to go out among the Chinese dead and scrounge for ammo. He told them to pick up any usable Chinese weapons, too—his best ballistics guys would clean and test-fire them so that at least some of them could be put to use. He got on the radio and requested resupplies by air.

Barber had barely recovered from the fighting, yet already he was preparing for the next night. He only had the daylight hours to organize his men. Once the skies darkened, the Chinese would be back.

=

Hector Cafferata's frozen feet never did connect with his abandoned boots. As he made his way toward them, shots rang out. His right arm was injured, and a bullet drilled into his chest, puncturing his right lung. He collapsed in the snow. He could scarcely breathe. His arm was in terrific pain—the radial nerve had been severed. Everything from his shoulder to his fingertips throbbed with a kind of hot electrical zing; it felt as though he'd taken hold of a high-voltage wire.

He screamed oaths at the sniper, whoever, wherever he was. Cafferata couldn't see anyone out there, couldn't even guess the general direction from which the round had been fired. He couldn't tell if it was from long range or from some shooter right next to him.

He figured he had it coming, that someone must have been paying attention and had taken account of the death spree for which this one burly American had been responsible. You didn't kill as many as a hundred men in a single night without karma catching up. Presumably the Chinese sniper had more to say—he wouldn't be content with merely winging his target. Cafferata felt completely exposed on the hill and dreaded what he knew was coming.

And several more rounds did come, whining overhead, flecking the snow around him. Luckily, they missed, and other Marines swooped in and poured fire into the area where the sniper seemed to be hiding.

"You okay, Hec?" one of them said between shots. "How bad is it?"

Other Marines started to crawl toward him, but Cafferata waved them away with his good arm. "I can make it!" he said. He didn't want them to expose themselves, too. He didn't think the wound was too serious. He only knew he had been hit in the arm; he wasn't aware that a bullet had entered his chest.

When Cafferata tried to stand, he couldn't catch his breath. He tried again, and this time he got to his feet, but his injured arm screamed with pain. It hurt so much he nearly fainted. He had to jury-rig a sling to carry the full weight of his ruined arm. He rummaged through his belongings, then got an idea. Using his teeth and the half-numb fingers of his left hand, he removed his ammunition bandolier, fastened it around his neck, and angled his arm through it. Leaving a trail of his blood as he went, he stumbled in his icy socks across the slope toward safety.

24

A HOT RECEPTION

Hagaru

At around ten thirty that same morning, November 28, General Smith climbed into the cockpit of a small Marine helicopter in Hamhung and took off for Hagaru—the place that would be his new headquarters. His mood was grave. That morning, he had been trolling the radio and gathering reports from his three regiments. The attacks at Yudam-ni and Toktong Pass were the most brutal, to be sure, but firefights had flared throughout the reservoir country. A large scattering of Army units encamped on the east side of the lake had been hit especially hard—reportedly suffering more than four hundred casualties.

As Smith's helicopter flew along the path of the MSR, it passed over nine separate Chinese roadblocks. The situation, he had to admit, looked "grim." On the way, his helicopter took some small-arms fire—nothing serious, but enough to make an impression. "These people were here to stay, and they were feeling cocky about it," said Smith's operations officer, Alpha Bowser. "Because they were so sure they could take us, they were no longer as concerned about staying out of sight. It was apparent—though no one had officially said so—that the Marines at Hagaru were in the process of being surrounded."

Smith was heading to Hagaru to take up residence in his newly prepared command post. Now that he had a true battle on his hands, he had little interest in leading it from the rear. He needed to be at the fulcrum of the fighting—and Hagaru was proving to be that spot, just as he had envisioned it might. At around eleven o'clock, Smith's helicopter hovered over Hagaru and crunched down in a frozen bean field. He jumped out into the bracing cold and tromped into a town that had become a maelstrom of activity. Hagaru had been spared the previous night, but recon patrols sent out into the surrounding hills had gathered intelligence that the Chinese were going to attack here in force on this night. Smith vowed, as one account put it, to give them "a hot reception."

The general set up his quarters in a two-room Japanese-built bungalow not far from the northern defense perimeter. The dank structure had a drafty stove and a lousy cot but little else to recommend it. It was a "pesthouse," said one account. "The air was close and foul, reeking with the musty odor of human bodies." Beside what would become Smith's desk, a large propaganda poster of Stalin sniggered from a grimy wall. When one of the staff members started to remove it, Smith, employing a bit of reverse psychology, decided it should stay. "Leave it be," he insisted. "Maybe it'll inspire us."

Smith first made sure he had ample quantities of his favorite tobacco—Sir Walter Raleigh—and then, filling and lighting his pipe, he wasted no time in getting to work. Incredibly, he hadn't yet received a single communication from X Corps in reaction to the previous night's withering attacks. "Apparently they were stunned," Smith said. "They just couldn't make up their minds that the Chinese had attacked in force. They just had to re-orient their thinking." But Smith knew instinctively that the march to the Yalu was over. His far-flung regiments couldn't defend themselves, let alone keep advancing. The enemy numbers were overwhelming.

For now, he ordered Litzenberg and Murray to stay put in Yudam-ni and await further instructions. "Until present situation clarifies, remain present position," read his terse message. "I halted the attack," he later explained, "because it was manifest that we were up against a massive force out there . . . We couldn't do anything but defend, as I couldn't withdraw without permission from higher authorities."

=

Acting independently, Smith had begun to conceive the broad outlines of a new battle plan. His goal was to consolidate his regiments in Hagaru and form what he hoped would be an impregnable legion there. As soon as possible, he would pull Litzenberg and Murray back from Yudam-ni to Hagaru and let them regroup there. The action would be something like a hen collecting her chicks under a protective wing. This ingathering of units, Smith knew, had to happen before anything else could seriously be entertained. Once pulled together, his Marines could then organize themselves for whatever was to come next.

And Smith already thought he knew what that was: a breakout, a fighting march back to Koto-ri—where his First Marine Regiment was now dug in—then down the many winding miles to the coastal plain around Hamhung. Here Smith imagined his division would spend the winter in bivouac, tending to its wounds and girding for a spring offensive. With its excellent harbor and airport and its favorably flat terrain, Hamhung seemed eminently defensible, Smith thought. It could be turned into a U.N. bulwark and held indefinitely—or at least until the powers that be, in Washington and in Tokyo, had formulated a new master plan.

But a fighting march from Hagaru down to Hamhung was not going to be easy. In military parlance, what Smith had in mind was a "retrograde maneuver"—also known as an "advance to the rear." Another word for it might be "retreat," although Smith assiduously avoided both the term and its connotations. Whatever euphemism one wanted to use, all the martial textbooks agreed on this point: Even under more favorable circumstances, a disciplined, well-choreographed fighting withdrawal was one of the trickiest maneuvers in military science (if anything having to do with the military could be said to be scientific). It was hard enough for an army to defend itself when it was dug in; to do so while on the move, with a numerically superior enemy attacking every inch of a rearward march, was next to impossible. Yet some battlefield situations offered only one solution beyond surrender or destruction—and that solution was a swift exit. The mark of a great general, Wellington once said, was "to know when to retreat, and have the courage to do it." Smith recognized that this was one of those times.

The secret to Smith's success, he knew, was going to be the airstrip. If Hagaru was the heart of the battle, then, to extend the metaphor, the airstrip would serve as both vena cava and aorta, bringing in and taking out everything Smith needed to keep his division alive and beating. Each day, on the wings of large airplanes, the vital supplies would come in, and the wounded would go out. The Hagaru perimeter would shrink; the division would distill to maximum efficiency. Finally, when the time was right, the Marines would break from this mountain citadel and bash their way to the sea.

Everything, then, was tied to the airfield. But Smith was miffed to learn how slowly the project was progressing. Though Lieutenant Colonel John Partridge's engineers had been slaving around the clock—working through the nights under the floodlights—they were only about 40 percent done. Partridge estimated that they had three days to go, maybe more. The dirt was frozen to a depth of eighteen inches. Even with sappers using explosives to soften the topsoil, it seemed to take forever for the dozers to scrape into this unforgiving earth. Smith would have to be patient.

A little later that morning, Smith was approached by Brigadier General Hank Hodes, the assistant commander of the Army's Seventh Division. Hodes was worried: He wanted to share with Smith the alarming reports he'd received from his Army units scattered along the east side of the reservoir. The GIs over there, under the command of Colonel Allan MacLean, had taken a serious hit the night before, with many hundreds dead and wounded. They were trapped, and Hodes didn't think they could fight their way out. General Hodes was too proud (personally and perhaps also institutionally) to say it out loud, but he hoped that Smith could help MacLean's men. He wanted the Marines to mount an expedition to go save the beleaguered Army troops. As Smith described it in his log, "The inference was that they should be rescued by a larger force." Alpha Bowser sensed that Hodes was "quite embarrassed about asking us for help in extricating the Army troops."

But the cruel fact—which Hodes must have understood—was that such a mission was impossible now. Smith didn't have enough men to defend Hagaru, let alone go marching up the road to rescue Army outfits stranded high along the reservoir. In truth, the unit that had been hardest-hit, the Thirty-first Regimental Combat Team, likely had twice as many troops in its ranks as Smith did in Hagaru. Until he could gather his regiments from Yudam-ni, Smith had precious little to work with here: a hodgepodge of three thousand Marines, along with a few hundred Army troops who happened to get caught here en route to the Yalu. Many of these potential defenders weren't fighters, at least not by formal job description. They were a medley of "casuals" and "service troops"—stenographers and clerks, signalmen and radio operators, cooks and MPs, carpenters and truck drivers. Whether generalists or specialists, they weren't exactly warriors. Many of them hadn't fired a weapon in years.

But so acute was Smith's need for warm bodies to plug the holes that every one of these service troops, Marine and Army alike, would be given rifles and sent out to the perimeter to fight with everyone else. They were "pretty much a bastard organization," one Marine colonel admitted—not a bunch "that would seem to inspire cockiness in any battalion commander facing a division of the enemy." But Smith's stronghold could not be allowed to fall; it would be defended to the last man, even if the last man was a baker or a mechanic.

In any case, Smith could not help the embattled Army units on the lake's east side (other than by offering close air support from the First Marine Air Wing). Said Alpha Bowser: "Any Marine force from Hagaru strong enough to blast its way through to the GIs would have left our perimeter dangerously vulnerable." Smith told Hodes, with some regret, that the Army battalions would have to fend for themselves. His recommendation was that they do everything they could to bull their way down the road to Hagaru.

≡

At midday, General Smith had an unexpected visitor. A small plane popped out of the leaden skies, circled overhead, and landed at Hagaru's old runway. It was an L-17 light aircraft, the *Blue Goose*. The plane's door sprung open, and out hopped General Ned Almond, crisply dressed in a warm parka, its collar lined with fur. Almond had a bounce in his step, and he carried himself with a cocksureness that didn't seem to align with the sober mood on the ground. It was as though he were trying to change the circumstances through force of body language, as though he thought that if he could telegraph sufficient confidence now, the present predicament might go away.

"General Almond preferred to believe that the reports of vast numbers of Chinese were grossly exaggerated, and that the distress signals from the reservoir represented little more than a loss of nerve," wrote historian Martin Russ. Almond had "decided to fly north and personally stiffen the collective backbone."

No account of their private meeting has been passed down, but the essence of it was apparent to Smith's staff: Almond had come to assure Smith, in the most emphatic terms, that X Corps was forging ahead, still bound for the Yalu. The previous night's attacks were unfortunate, but they wouldn't stop the campaign. Smith's division was to keep moving, as though nothing had happened.

Near the end of the conversation, Smith's aides watched the two generals standing off by themselves, Almond holding forth, gesticulating with bravado. Smith listened with a look of disbelief. He made a dismissive gesture, then spun around and stormed off. As he passed by one of his aides, Smith muttered through gritted teeth, "That man must be crazy."

A few minutes later, General Almond crawled into a Marine helicopter and buzzed over to the east side of the reservoir, to get a quick look at the besieged Army units that had been so ferociously attacked the previous night. He sped up the eastern shore, over the weave of icy inlets and coves. Ten minutes later, the chopper landed in a field by the road, and Almond marched over to the command post of Colonel MacLean, commander of the Thirty-first Infantry Regiment. Dead bodies, Chinese and American alike, littered the surrounding hills.

Almond spread a crinkly map over the warm hood of a jeep and reviewed the previous night's fighting with MacLean's immediate subordinate, Lieutenant Colonel Don Carlos Faith. Faith was a golden boy, a graduate of Georgetown University and the son of a general—a handsome young man, and brave to boot. Faith, who had served in China during World War II, had a good grasp of the battle situation here, and in his view it was bleak. He estimated that two Chinese divisions had fallen upon the Thirty-first Infantry the night before.

"That's impossible," Almond countered. "There aren't two Chinese divisions in the whole of North Korea!"

Faith tried to press his view of things, but Almond smothered him. "I want you to retake the high ground you lost during the night," he insisted. Almond railed at the tentative behavior he saw around him, all this worry and deliberation and timidity. This was not how an American fighting force was supposed to face what he considered a crude bamboo army, a medieval clan of rice farmers. "The enemy delaying you," Almond told Faith, "is nothing more than the remnant of units fleeing north. We're still attacking, and we're going all the way to the Yalu." Then Almond uttered a line that would live in infamy. He said, "Don't let a bunch of goddamn Chinese laundrymen stop you!"

Maybe Almond felt bad for castigating this resolute young man, Faith. Or maybe he was instinctively trying to mimic his own boss, General MacArthur, who had a soft spot for ceremony. But what Almond did next was a priceless bit of misplaced theater. He decided it was time to award Faith a medal—here and now, with corpses lying around them, in the midst of a crisis. Almond had a Silver Star in his pocket. After running through the usual military rigmarole, he pinned the medal on Colonel Faith's parka.

"Now you retake that hill," Almond reiterated. Then he turned and bounded back to the idling helicopter.

"What a damned travesty," Faith said under his breath. He was disgusted by the whole charade. As the helicopter rose into the sky, Faith ripped his Silver Star from his parka and tossed it into the snow.

≡

Almond made his way back to Hagaru and that afternoon took off in the *Blue Goose*, climbing away from the reservoir country. Most people in Hagaru were happy to see him go—General Smith especially. He could now spend the rest of the afternoon without further distraction, getting ready for the fight he knew was coming that night. In leaving so quickly, Almond seemed to have sensed that Chosin was Smith's realm now, Smith's problem to solve. On some level, the X Corps commander must have also sensed that he was neither liked nor needed here.

Smith sucked at his pipe in his dungeony quarters and studied the topographical maps of Hagaru. He was especially concerned about fortifying the new airstrip. He assumed that the Chinese would do everything in their power to attack it, hoping to destroy the five Caterpillar tractors that had been clawing at the frozen turf, night and day. He would put some of his best companies there. Another point of concern was East Hill, a promontory that loomed over the edge of Hagaru. Doubtless the enemy would try to seize this strategic eminence, which afforded a commanding view of the town. Smith didn't have enough regular infantrymen to defend it; he knew his army of bakers and clerks would be critically needed to flush the Chinese from East Hill's slopes.

Through the afternoon, along the perimeter, the men readied themselves for battle. They strung strands of barbed wire. They blasted foxholes with putties of C3 explosive. They planted mines and booby traps. They loosened pins on grenades and taped them to stakes fastened to trip wires. They sharpened their bayonets and cleaned their rifles, wiping off the Cosmoline and other oils that tended to congeal in the freezing cold. Possibly every single person in Hagaru was armed now—even General Smith, who had a .45 sidearm tucked away in a holster.

In the mess tents, the cooks ladled hot chow in rotations and put out five-gallon vats of fresh coffee. Everyone took their positions around Hagaru and settled down for the night. A 50 percent alert order was upgraded to 100 percent. A fine, powdery snow began to fall, coating everything, dampening the miscellaneous noises of the valley. Then the night grew silent. The snowflakes made tiny ticking sounds as they accumulated on rifle barrels. Everyone could sense that the Chinese were close at hand.

Then they appeared—as they had the previous night at Yudam-ni and Toktong Pass. Here came the smashing cymbals, the blatting bugles, the shrilling whistles—an "exotic concert," as one Marine put it. And then masses of men, moving through the night, throwing themselves at the Americans. Said one Marine: "It was as though a whole field got up on its feet and walked forward. I never saw anything like it." Said another: "A hell of a lot of Chinese went down, but a hell of a lot more kept coming. You got the impression the waves were endless, like surf lapping on a beach."

The cooks and the bakers, the stenographers and the clerks, nobly acquitted themselves on East Hill. And the engineers at the airstrip tossed aside their implements and crawled down from their earth-moving machines to pick up rifles and drive the Chinese away.

Smith seemed calm throughout the fighting, suffused in the blue smoke of Sir Walter Raleigh, working beneath the gaze of Joseph Stalin. This was what he lived for—combat on a grand scale, a complicated panorama of interlocking details. In a letter he later wrote to his wife, he rhapsodized about the "chattering of machine guns, the cough of mortars, and the booming of artillery." During the night, bullets perforated his bungalow walls. Smith described "unusual sound effects" as the unwanted missiles plunked off the galley pans, creating a "witch's clatter," as one account described it. Smith was thrilled by it all. He had vowed to give the Chinese a hot reception, and that was what he'd done.

25

THE WAR COUNCIL

Tokyo

On the afternoon of the twenty-eighth, upon returning to Hamhung from his visit to the reservoir high country, General Ned Almond received an urgent summons from MacArthur's headquarters. He was to fly at once to Tokyo for an emergency meeting. The message was marked top secret. Almond was to tell no one.

The X Corps commander went straightaway to Hamhung's Yonpo Airfield and boarded a C-54 Skymaster with a few close aides. At nine thirty that night, the plane arrived at Haneda Army Air Base, where Almond was met by a Far East Command colonel who escorted him to MacArthur's residence in the American embassy.

The occasion for the summons immediately became clear: MacArthur had assembled a war council, a conclave of his most reliable chieftains from the Army, the Navy, and the Air Force, as well as close aides from his immediate staff. Not only had MacArthur summoned Almond from Korea; he had also called General Walton Walker, whose Eighth Army, on the peninsula's west side, had been thoroughly surprised and overwhelmed by another massive force of the Red Chinese, and were now in headlong retreat.

There was no mistaking the dolorous mood that had descended over MacArthur's world in Tokyo. "This headquarters is in a terrible slump," wrote intelligence officer Lieutenant Colonel James Polk. "When a gambler pulls one off he is hailed as a genius, and when he fails, he is a bum. This time [MacArthur] failed, and he has to take the consequences. Per always, he made his mind up all by himself, he gambled on his intuition, and the plain fact is that he lost." Now, said Polk, "this whole house of cards [will] suddenly fall down with a resounding smash."

The reports that had filtered into MacArthur's headquarters over the past twenty-four hours had left the supreme commander in a nearly catatonic state. It was as though he were frozen, unable to think or act. He was moody, irritable, depressed—in a blue funk, it was said. He paced the floor. He couldn't sleep or eat. He couldn't believe this turn of events—to have been checkmated by a primitive army of serfs in tennis shoes.

MacArthur was already lashing out in various directions—first at Mao and the Red Chinese, who had double-crossed him, hood-winked him, broken the conventions of civilized warfare. Their entry, he said, amounted to one of the most treacherous sneak attacks in history, worse even than Pearl Harbor.

The supreme commander lashed out at Washington as well—at Truman and Acheson, at the Joint Chiefs of Staff. At every turn, he contended, they had hamstrung him. If they had only let him bomb the Chinese in Manchuria, bomb their bases and roads and installations, this debacle would never have occurred. The appeasers in Washington, with their fatuous concerns about keeping the war "limited"—*they* were the ones who had made this happen.

If it is true that a fish rots from the head down, then MacArthur did not know it. His organization in Tokyo, which had put so many American troops in harm's way, was a nearly perfect reflection of himself. Yet the man was incapable of accepting blame, or assuming responsibility, for the mistakes that had been made. Already he was beginning to cover his tracks, to write his own posterity papers. He had started to formulate a defense for himself, a counter-narrative that, in many ways, would appear to be delusional. He would argue that he had known all along that the Chinese were going to intervene en masse. He had seen it coming for many weeks.

MacArthur claimed that he had sent his armies north expressly to learn the size and disposition, the content and intent, of the enemy troops. His advance had accomplished precisely what he had planned from the start: It had triggered a Chinese assault, had forced them to show their hand. Charles Willoughby, MacArthur's intelligence chief, articulated the spurious thinking this way: "We couldn't just passively sit by. We had to attack and find out the enemy's profile."

MacArthur had "not been taken by surprise," insisted his close aide General Courtney Whitney. "His troops did not rush blindly north into a massive ambush. The push north had been carefully designed . . . as a reconnaissance in force"—and, as far as Whitney was concerned, it had worked brilliantly.

The sophistry emanating from Tokyo was dazzling in its desperation. MacArthur, wrote journalist and historian David Halberstam, had "lost face not just before the entire world, but before his own troops, and perhaps most important of all, before himself." And so, to preserve the illusion of omniscience, MacArthur had spun an elaborate retroactive fiction. *Reconnaissance in force*—that was Tokyo's new incantation.

≡

At the war council, such talk was ambient noise, static in the air, bile festering in the bowels of the Dai Ichi Building. These were early versions of arguments and recriminations that would be sent across phone lines and telex wires in the weeks ahead. The real reason MacArthur had summoned General Almond and General Walker on this night was that he wanted to hear from them. He wanted to know what they had seen on the battlefield, that very day. It was a rare moment for MacArthur. He was genuinely asking them: What did *they* think? What did *they* know? What would *they* do?

General Walker was the first to speak. His Eighth Army was now withdrawing toward Pyongyang. He was worried that the Chinese would outflank him. He thought he could hold a line somewhere around Pyongyang, but he wasn't sure. The situation on the west side of the peninsula looked hopeless. Some of his U.N. units were performing valiantly—especially a brigade of Turks, who had surprised everyone with their resilience and ferocity. But overall, the west side was devolving into a rout.

Almond, predictably, was feistier. He didn't want to abandon the grand plan. He thought his X Corps could keep going. His men had suffered momentary setbacks, but they would recover, and they would carry on to the Yalu. And in doing so, they would sever the Chinese lines of supply and communication. Like an overeager acolyte, Almond still held MacArthur's torch.

While no written record of the meeting exists, what is known is that MacArthur categorically rejected Almond's sanguine assessment of the battle. The supreme commander's earnest lieutenant could be forgiven for his zeal, but the campaign to reach the Yalu was over. Even MacArthur understood that. It was time to switch over from the offensive to the defensive. MacArthur told Almond that the new strategy was for his X Corps to concentrate in the plains around Hamhung-Hungnam. Maybe they could hold on to the twin cities for the winter, or maybe they would have to quickly evacuate, with Navy ships taking advantage of Hungnam's deepwater port. That question could be decided later. The bigger point was this: General Douglas MacArthur had finally come to see the futility of the advance. The mirage had evaporated; the dream had been extinguished. The Home for Christmas campaign had shriveled down to a dire exercise in self-preservation. X Corps was turning around and marching for the sea.

In the morning, a chastened Almond climbed back onto a plane at Haneda Airport and returned to a different battlefield.

26

AN ENTIRELY NEW WAR

Washington, D.C.

At six fifteen on the morning of November 28, President Truman received a disturbing phone call at Blair House from General Omar Bradley, the chairman of the Joint Chiefs of Staff. MacArthur had cabled earlier that morning with what Bradley called a "terrible message."

"We face an entirely new war," MacArthur's cable had said. "This command has done everything humanly possible within its capabilities but is now faced with conditions beyond its control and its strength."

"What's the estimate of the Chinese troop strength?" Truman pressed.

Bradley was somber. "MacArthur now believes it's 260,000 men." He said MacArthur would have to go on the defensive. He could no longer contain Mao's attack; he could only try to escape from the trap the enemy had set. The Chinese, MacArthur said, were bent on the "complete destruction" of his forces. Not only were the Marines under attack at the Chosin Reservoir; the Chinese had also massively assaulted General Walker's troops on the west side of the peninsula. His Eighth Army and assorted other U.N. forces under his command were scurrying southward.

What a stunning reversal this was for the president. Only three days earlier, MacArthur had declared his "end-the-war" offensive. Now, in a jolting moment, the situation in Korea had been transformed, in Truman's words, from "rumors of resistance to certainty of defeat."

Thus began one of his darkest days in office. Four hours later, at his staff meeting in the Oval Office, Truman broached the news with his closest aides. "We have a terrific situation on our hands," he said. "The Chinese have come in with both feet." He mentioned the astonishing number MacArthur had passed on to Bradley.

One of his staff members couldn't believe what he'd heard. *Two hundred sixty thousand?*

"That's right," the president said. "They have something like seven armies in there."

His aides sat in stunned silence while Truman, his eyes magnified behind his thick glasses, struggled with his emotions. "His mouth drew tight, his cheeks flushed," recalled one observer in the room. "For a moment, it almost seemed as if he would sob." The president awkwardly shuffled papers and fiddled with a pair of scissors on his desk.

But then he regained his equilibrium. "This is the worst situation we have had yet," Truman said. "We'll just have to meet it like all the rest." He seemed to find solace in the making of concrete plans. He would declare a state of emergency. He would deliver a speech to the nation. He would triple the Pentagon's budget. "I'll have to ask you all to go to work and make the necessary preparations," he said. His appointments, on that day and the days ahead, would need to be canceled.

"We have got to meet this thing," he repeated. "Let's go ahead now and do our jobs as best we can."

≡

At three o'clock that same day, November 28, Truman met with the National Security Council in the Cabinet Room of the White House's West Wing. The mood around the table was funereal. Omar Bradley shared his dour thoughts, as did George Marshall, Vice President Alben Barkley, Air Force Chief of Staff Hoyt Vandenberg, and CIA director Walter Bedell Smith.

Then it was Dean Acheson's turn to speak. The magisterially mustached secretary of state was Truman's most influential adviser.

Elegant, eloquent, and arrogant, Acheson was a kind of super-WASP: Connecticut-born; son of an Episcopal bishop; product of Groton, Yale, and Harvard Law; an international attorney before rising into the inner sanctums of government. Critics found the debonair Acheson criminally out of touch—"an ancient mandarin," thought one prominent newspaper columnist, "frigid, aloof . . . untempered by emotion," though sometimes "forced by the vulgar circumstances of life to speak to those who never could understand the world of great affairs."

Acheson's State Department had been embroiled in numerous controversies—most of them having to do with China, or with threats, real and imagined, concerning the spread of Communism. In a sense, Red China's entrance into the war was only the newest wrinkle in a much larger saga that had preoccupied his days at Foggy Bottom. In 1949, when Mao's forces prevailed over Chiang Kai-shek, hawkish conservatives widely castigated Acheson for being soft on Communism. Many Republicans, labeling Acheson an appeaser, personally blamed him for having "lost" China. When Kim Il Sung invaded South Korea, Truman administration critics recalled that Acheson, in an important speech he'd given at the National Press Club back in January, had failed to single out Korea as one of the Asian nations the United States was committed to protecting under its broader security umbrella. Some asserted that Acheson's omission had so emboldened Kim that it may well have been one of the precipitating factors of the invasion. Once Kim attacked, it was Acheson, against the recommendation of many other Truman advisers, who had convinced the president to commit troops to the Korean Peninsula—scarcely guessing that this fateful decision would lead the United States into war with China.

Earlier that day, Acheson had given a sobering account of the Korean situation to an executive session of the Senate Foreign Relations Committee. "I think it is impossible to overestimate the seriousness of this whole matter," he said. "Not merely the immediate military situation in Korea, but what it *means*." He said the United States was "very close" to world war. "We have got to face the possibility now that anything can happen anywhere at any time."

Now, speaking in the Cabinet Room, drawing out his lockjaw locutions, Acheson struck the same apocalyptic tone. The president listened, riveted, blinking like an owl. With China's incursion into the war, Acheson said he perceived the hidden hand of Stalin. "We must consider Korea not in isolation," he warned, "but in the worldwide problem of confronting the Soviet Union as an antagonist."

Acheson had trouble understanding how MacArthur had put U.N. forces in such a position, how his intelligence had been so wrong, and how he had delayed for so long before honestly confronting the situation he faced. The supreme commander had come close to digging "a hole," Acheson said, "without an exit." The secretary of state thought that, somewhere along the way, MacArthur had lost touch with reality—he seemed hysterical at times, even delusional. (In describing him, Acheson would later quote Euripides: "Whom the gods would destroy they first make mad.")

Acheson said he would devote careful thought to how the United States might punish China for its actions. "We should see what pressures we can put on the Chinese Communists to make life harder for them," he said. But at the same time, China was not a country that could be defeated. They would keep throwing more and more men across the Yalu—millions of them, if necessary. Mao placed scant value on human life. Said Acheson: "We can't beat China in Korea. They can put in more than we can. Our one imperative step is to find a line that we can hold, and hold it."

Then, he said, "we must terminate the fighting, turn over some area to the Republic of Korea—and get out."

=

The full and frantic day had brought much for President Truman to ponder. He had to agree with Secretary Acheson: MacArthur had blundered badly. He had been outwitted and outflanked by a guerrilla army with no air force, crude logistics, and primitive communications, an army with no tanks and precious little artillery. He was responsible for one of the most egregious intelligence failures in American military history. Somehow he had missed the significance of a monthlong accretion of evidence—evidence that, when he finally recognized its import, he had either suppressed or ignored. In the process, he had put many tens of thousands of American lives in mortal danger. "I should have relieved General MacArthur then and there," Truman wrote. "The reason I did not was that I did not wish to have it appear as if he were being relieved because the offensive failed. I have never believed in going back on people when luck is against them, and I did not intend to do it now."

Increasingly, Truman had begun to see that Korea was only part of a much larger peril that his administration faced—one that had been summed up in a top-secret policy paper, called NSC-68, that the president's staff had been revising over the past few weeks and months. Written by a committee of policy mavens under the direction of Paul Nitze, NSC-68 was a sweeping analysis that would become one of the most influential documents in American history. It would guide and define United States foreign policy for the next twenty years. Seeing the containment of Soviet expansionism and the hydra-headed Communist threat as the paramount concern of American statecraft, the highly classified fifty-eight-page paper advocated, among other things, a dramatic increase in the military budget of the United States, the development of the hydrogen bomb, and expanded military aid to allies, satellites, and puppet states around the world. Above all, NSC-68 captured the dire tone of the times: "The issues that face us are momentous, involving the fulfillment or destruction not only of this Republic but of civilization itself."

Margaret Truman would later point to this day as the start of one of the "grimmest" times in her father's tenure. Truman had presided over the end of World War II. He had helped to rebuild Europe with the creation of the Marshall Plan and NATO. He had stood down Soviet aggression with the Berlin Airlift and checked the rise of Communist movements in Greece and Turkey. But this, he thought, was his steepest challenge yet.

"It looks like World War III is here," he wrote. "I hope not—but we must meet whatever comes—and we will."

BOOK
FOUR

RED SNOW

Go, way-farer, bear news to Sparta's town
That here, their bidding done, we laid us down.
—CYRIL E. ROBINSON

27

YOU WILL ALL BE SLAUGHTERED

Toktong Pass

Hector Cafferata, staggering in his stocking feet and dripping blood on the snow, finally made it down the hill to Fox Company headquarters. He found the medical aid tent, which had been set up in a stand of pines. Cafferata's arm, in its makeshift sling, throbbed with pain, and he was having trouble breathing. His feet were swollen and icy. Inside the canvas tent, the scene was primitive: rows of badly wounded men curled on the bare ground, which was strewn with plasma bottles and morphine syrettes. The tight space stank of sweat and shit and the foul odors of exposed guts. Some of the less seriously wounded passed around a bottle of White Horse Scotch to relieve the pain. Just behind the tent lay the frozen stacks of the American dead.

At least it was halfway warm inside. A kerosene stove threw off welcome waves of fumy heat. Fox Company had no doctors, but three Navy corpsmen bustled about the tent, tending to the night's casualties. One of the corpsmen, James French, guided Cafferata to the triage section to get him stabilized. French cut away Cafferata's bloody sleeve, cleaned and dressed his hand and arm wounds, and gave him a shot of morphine. "Holy Christ, that was a weird feeling," Cafferata said. "It seemed like in seconds my arm left my body. Or maybe it was my body that left my arm."

The now pleasantly medicated Cafferata found his platoon mate Ken Benson, who was still effectively blind from the blood and debris crusted in his eyes. Benson, fumbling a bit, got Cafferata's icy socks off and rubbed his feet to try to coax some life back into them. The two shivering New Jersey boys shared a canteen cup of hot coffee and huddled for a while in the tent. At times Cafferata, in his narcotic haze, made little sense. He wondered what had happened to his boots. He wondered what had happened to the three Chinese prisoners he had taken. He wondered how he had survived the night. "I don't know why I'm here," he said. "Don't know how the hell I didn't buy the farm. I guess somebody was lookin' out for me."

Around midday, Benson had to say goodbye. The corpsmen had rinsed his eyes and had carefully tweezed out the grit and the shards and the clots of blood. His glasses had been shattered on the hill, but now he could see well enough without them to be sent back up to dig in and ready himself for the enemy's return. Cafferata was another matter. He could barely move, and he began to sense that the pain wasn't just in his ruined arm. It was somewhere deep in his upper torso—he felt as if he had a case of double pneumonia. He couldn't understand why—there was no blood seeping from his chest, no wound that he could find. French and the other corpsmen were too busy dealing with more serious cases. And so, through the afternoon of the twenty-eighth, Cafferata could only lie back and listen to the cries and moans of the wounded.

≡

Captain Bill Barber had been on the radio much of the day, trying to learn whether reinforcements could be brought in to Fox Hill. He had trouble getting a signal—the batteries in his radio had been drained by the cold. He knew the Chinese would be back after nightfall. Not that they had really gone anywhere: He could see them in the copses and thickets, lingering just out of range. They seemed to be taunting his Marines. If not for American air superiority, the Chinese would be attacking even now.

Against such odds, Barber wasn't sure how much longer he could hold. On the radio, there was talk of a relief party marching down from Yudam-ni, and talk of another relief party marching up from Hagaru, this one supposedly composed primarily of Marine cooks and bakers. But those plans had quickly been scuttled: The road was simply too dangerous. The Chinese had seized control of it and erected numerous roadblocks. Fox Company was on its own.

Barber summoned his officers and tried to explain the situation. "There's no possible way we can be relieved right now," he said. Litzenberg and Murray had all they could handle at Yudam-ni. General Smith had his hands full at Hagaru, too. It seemed that the First Marine Division was surrounded by as many as four or five divisions of the Red Chinese Ninth Army Group. For the foreseeable future, the men of Fox Company would have to fend for themselves.

"We can expect heavy attacks again tonight," Barber warned. "But we have nothing to worry about, as long as we fight like Marines."

The captain had some good news, however. He had been promised an aerial drop of ammunition and other crucial supplies. Sure enough, at around three in the afternoon, a Marine cargo plane came roaring over Fox Hill. The men on the ground let out lusty cheers as the hatch opened and a number of pallets emerged from the plane's belly, dropping by parachute to the ground. Unfortunately, the load crashed in no-man's-land, a good seventy-five yards outside the Fox Company perimeter.

Supply sergeant David Smith was the first to reach the pallets. He took out a knife and was beginning to cut the parachute cords when a Chinese sniper fired at him. The bullet struck Smith's right calf, shattering his tibia, making a noise that sounded like the snap of a dry branch. Smith yowled in pain and collapsed into a ditch. A team of men ran out with a stretcher to rescue Smith, but one of them, too, was hit in the leg by a sniper. Machine gunners then laid down a fusillade of covering fire as a larger group of Marines fetched the two crippled men and recovered the supplies.

Captain Barber was more than pleased with the trove: He found grenades and mortar rounds and illumination shells and numerous boxes of .30 caliber ammunition. There were also blankets and stretchers and other medical supplies needed by the corpsmen. Even the parachutes would be put to good use—the silk would be ripped into strips to serve as clothing or bedding material.

Not long after the drop, the men of Fox Hill were surprised to be visited by yet another aircraft: a tiny two-seater helicopter. This was high country for choppers, a newfangled craft dangerously unreliable in thin air, especially in such cold weather. The pilot of this helicopter, Captain George Farish, was a brave soul. He had come from Hagaru to deliver fresh batteries for Barber's field telephones and radios. Farish came buzzing over the treetops and hovered for a moment above Fox Hill, looking for a good place to land. But then a Chinese sharpshooter fired on him. One of the bullets tore into the chopper's rotor transmission case, and it began to spew oil.

Farish was in serious trouble. He was losing control of the chopper—it careened and wobbled in the air. He gave a little salute, and as he tried to climb away, the rotor blades sawed at the tree branches. Farish managed to fly back to Hagaru, only to crash on the outskirts of town. Miraculously, he wasn't seriously hurt—he climbed out of the mangled chopper and walked the rest of the way back to General Smith's compound.

≡

That afternoon, Ken Benson traipsed back to the place where he and Cafferata had made their stand the night before, beside the scattering of boulders. He scrounged for any useful items they may have left behind. He found their sleeping bags, but they were a mess—the Chinese had shot them and stabbed them with their bayonets, presumably thinking that Marines were snoozing inside.

Benson, watching the tufts of down and the tangled fibers rustle in the wind, realized how lucky he was to be alive.

But then he had an idea. Why not take the two sleeping bags, and any others that weren't being used, and stuff them with snow? He could place the bags strategically in a fantail pattern on the hill, to make it look as though a Marine platoon was encamped together. These "sleeping" dummies would serve as perfect decoys. They would draw enemy fire, and the muzzle flashes would give away the Chinese positions.

The Marines, outnumbered as they were, had to be clever like this. Ruses and tricks were going to see them through their ordeal.

Benson's platoon sergeant liked his idea. After nightfall, Benson got to work, making snowmen and arranging the decoys on the hill.

=

That night at around ten, the men of Fox Company were roused by an eerie noise. They heard what sounded like electronic static, then a din of feedback. Then they heard a man's voice, amplified through a loudspeaker. Though it was a Chinese voice, the orator spoke beautiful English, tinged with a British accent. He enunciated his words with pedantic precision. "Fox Company," the man announced. "You are completely surrounded! You are greatly outnumbered!"

The Marines craned their necks to see where the amplifier was. A few of the men, equipped with field glasses, caught a glimpse of the Chinese orator. His face danced in the light of a bonfire. He looked to be an imposing man, wearing an officer's cap and a quilted greatcoat. "Marines of Fox Company," he went on. "You must know that the only rational course is surrender. Otherwise, you will all be slaughtered!"

The voice trailed off, and the trigger-happy Marines resisted the instinct to pump automatic fire in the direction of the hated voice. Then the strains of a familiar song, sung by a familiar voice, began to issue from the loudspeaker. "Where the treetops glisten and children listen to hear sleigh bells in the snow . . ." It was Bing Crosby, his sonorous baritone carrying on the wind, seeping through the trees, stabbing the men with homesickness.

Moments later, another Chinese voice broke the reverie. The speaker bellowed in pidgin English, chanting over and over again: "Marines, tonight you die! Marines, tonight you die!"

=

Around midnight, the onslaught began, much as it had the evening before. Only this time, the Marines were ready. The Chinese whistles and horns and cymbals did not seem so frightening this second night. The men of Fox Company, having tightened their perimeter and dug their foxholes deeper, did not waste their bullets. They waited patiently until the advancing figures stole into plain view. Then the Marines cut them down.

By the light of a flare, many of the men caught sight of a lone bugler on the hill. The Chinese man stood in perfect dignity as he brought his trumpet to his lips. "He was motionless," Marine Robert Leckie wrote, "a heroic figure out of an antiquity when Mongol ponies trod the earth of Europe and yak-tails swung at the tents of the Golden Camp." The bugler played a lugubrious note that resonated over the hillside. "Lemme fix the bastard!" one of the Marines shouted. He unpinned a grenade and threw it, and it landed at the bugler's feet. But the dignified man didn't run; he just stood there, implacable, blowing his horn as the seconds ticked off. Then the grenade exploded. Wrote Leckie: "There was a flash and a roar, intermixed with the long trailing wail of a horn—and the din of battle reclaimed all ears."

Down the perimeter line, Ken Benson, vigilant in his foxhole, waited patiently for his ruse to pay off. It wasn't long before a squad of Chinese came streaming up the hill. They reached the crescent of bulging sleeping bags Benson had prepared for them. The Red soldiers hesitated for a moment. Then they aimed their rifles and fired into the bags at point-blank range. The light of their muzzles gave them away. Benson and his comrades easily sighted in on the Chinese—and killed them all.

The fighting continued through the night and into the morning hours. The whole time, Captain Barber kept to the front lines, urging on his platoons as he had the night before. The man was fearless. He took crazy chances, he got right in the fray—just as he had at Iwo Jima. In some ways, he'd never been happier. Barber was a natural at this—not only a warrior, but a leader of warriors.

But around 2:45, his luck ran out. The man who'd said, "They haven't made the bullet yet that can kill me" was struck hard. A round from somewhere drove deep into his groin and shattered his pelvis. A red splotch bloomed high along his left thigh. At first, Barber thought little of it—he said it wasn't much more than a bee sting. He jammed a swatch of cloth into the wound. He broke off a stick to use as a crutch and tried to move around. When someone called for a corpsman, Barber scoffed. He was so flushed with adrenaline, it took him some time to settle down. When he did, he saw that he was useless. Now the pain was excruciating. The captain agreed to report to the aid tent, but he would walk on his own power: He wouldn't accept a stretcher. Barber found another wounded officer to lean on, and the two men lurched along the hill, trying to hold each other up.

From somewhere in the darkness, a Chinese voice cried out: "Captain Barber, will you surrender?" He would not.

≡

As the combat raged outside, the wounded men in the medical tent were antsy and restless—Hector Cafferata especially. He felt he should be out there, fighting alongside Benson. He could hear the gunfire and the mortars, could hear the shouts of the men. Every so often, a bullet would rip through the canvas tent, clinking against the heater, letting in a draft of cold air or a wink of starlight.

Captain Barber came clomping in, wincing but pumped up with battle fervor. "We're short of warm bodies," he barked. "We need volunteers to plug the lines." Several men who were not too seriously wounded answered the call—they struggled to their feet and marched outside. Others tried to stand but couldn't. One of them, Private Harrison Pomers, complained about a wet, tacky sensation somewhere along his back. A corpsman peeled away his clothing and gasped: Pomers had a hole as big as a fist. The corpsman could see his spinal column.

Now it was Cafferata's turn to heed Barber's call. He tried to pull himself off the ground. He yearned to join the battle. Earlier in the day, Benson had given him a Mauser machine pistol, and Cafferata intended to use it.

But French, the corpsman, saw him struggling. He didn't think Cafferata should go. "You lay down, Moose!" he said.

Cafferata protested: "Frenchy, I should be out there." He didn't understand how a gunshot wound to the arm had sidelined him when so many other men out on the hill were fighting with far more serious injuries.

But in truth, Cafferata couldn't move. The pain in his chest had only grown more severe through the day, even with the morphine French had given him. Seeing how much distress he was in, French pulled back Cafferata's clothes and studied his chest in the guttering light of a candle. Just below the sternum was a tiny red bullet hole. It was an almost immaculate wound, bloodless and clear. Somehow they'd missed it all day.

"You damn fool!" French snapped. "Lie down!"

This time, Cafferata relented. He could see by French's eyes that this was serious. French didn't have any way of treating it. He could only give Cafferata another shot of morphine. Cafferata settled back on the ground and, wrapping himself in a blanket, tried to digest the news. He'd been shot in the chest—*that* sounded like an honorable battle wound. It picked up his spirits. At least he no longer felt bashful about lying around in a medical tent.

As the terrors of war blasted around him, Cafferata gripped his Mauser and stared into the darkness, alone with his thoughts.

28

KISSING A BUZZ SAW

New York City

Four days passed before the Chinese ambassador, Wu Xiuquan, was cleared to speak at the United Nations. But in those four days, the world had changed. Mao's intentions in Korea had been fully revealed. Now the forces of the United States—and of the United Nations—were in a state of actual war with China, surrounded by hundreds of thousands of Communist Chinese soldiers in the field. Whatever slender hopes still existed that Wu might be willing to negotiate had shriveled.

If anything, Mao's early battlefield victories only emboldened him, and Wu was instructed to take an especially hard line in his dealings at the U.N. From their temporary offices in the Waldorf-Astoria, Wu and his fellow delegates stayed in nearly constant contact with Zhou Enlai and his ministers in Beijing. Zhou was euphoric over the news from Korea. It seemed the Chinese were on the verge of destroying the Americans—which, to Zhou, meant the Chinese could present their arguments to the world from a position of strength.

Finally, on November 28, Ambassador Wu was slated to address the United Nations Security Council, a body that did not recognize him or his government but had invited him to speak just the same. The United Nations, its Manhattan headquarters under construction, was temporarily housed in a cavernous building formerly run by the Sperry Gyroscope Corporation, in a suburb of New York called Lake Success, on Long Island.

The Chinese, wearing black, excited much interest and elicited stares as they marched into a committee session already in progress—

some delegates thought they'd strutted in like gangsters. Coincidentally, the speaker holding the floor was Andrei Vyshinsky, chief of
the Soviet delegation and the notorious prosecutor at many of Stalin's
purge trials. Vyshinsky paused to give the Chinese an effusive welcome. Wu took a seat at a desk beside a placard that read PEOPLE'S
REPUBLIC OF CHINA. Two seats over sat the poker-faced John Foster
Dulles, special counsel to the State Department. Though Dulles pretended not to notice the new arrivals, Wu glowered at him and held
his gaze for several uncomfortable minutes.

Later, in a packed hall before the Security Council, Ambassador Wu removed his earphones and began his formal speech. Standing over his notes, speaking in staccato bursts, his voice piercing and
shrill, he wasted no time in skewering the host nation. "I am here in
the name of the 475 million people of China to accuse the United
States government of the unlawful and criminal act of armed aggression. It is an integral part of the overall plan of the United States to
intensify its aggression, control, and enslavement of Asian countries,
a further step in the development of interference by American imperialism in the affairs of Asia. The American imperialists have always
in their relations with China been the cunning aggressor."

The speech went on in this vein for the better part of two
hours—an extended harangue, twenty thousand words in length.
The tension in the hall was palpable, and listeners squirmed restlessly
in their seats. Wu did not seem to offer a single conciliatory note. He
blurted his words faster than the U.N. linguists could translate. His
demeanor was combative, and his facial scar seemed to glisten under
the klieg lights.

Every word of the speech, it was later revealed, had been written
by Zhou Enlai's ministers in Beijing, with direct consultation from
Mao himself. The U.N. delegates could see that the Chinese were not
disposed to seek a diplomatic solution at all. On the contrary, they
seemed to wish only to escalate the conflict. They demanded that
the United States immediately leave Taiwan, leave Korea—leave Asia
altogether.

Later in the speech, Wu began to give voice to a deep-seated animus that he said China had long felt toward the United States. It was as though a century's storehouse of affronts and humiliations had come bursting forth from China's collective memory and now poured through the headsets in a babble of languages. In Wu's telling, the United States had always been at war with China. Sometimes it took the form of a cultural or trade war, sometimes a military one, but the United States had an incorrigible habit of insinuating itself into China's domestic affairs. American interference during the Boxer Rebellion, the patronizing slights of American missionaries, the insults of Yankee gunboat diplomacy, Washington's support of Chiang Kai-shek, and now the U.S. Navy in the Strait of Taiwan and American ground forces driving toward Manchuria—all of it was just more evidence of a sly and meddlesome aggressor. "The real intention of the U.S.," Wu railed, "is to dominate every Asiatic port from Vladivostok to Singapore."

The crisis in Korea turned out to be a secondary complaint within Wu's address. "The main burden of my speech," he later said, "concerned all the wrongs China had suffered at the hands of the Americans. I was more anxious to display our country's indignation."

Wu's long rant, turgid with Marxist terminology, shocked many in the hall. It seemed to some listeners that Communist China, having been offered a world stage and a megaphone for the first time, was behaving like a churlish teenager. Wu expressed no remorse for the fact that his country's armies had attacked the military forces of this body—in fact, he didn't even mention it. He had come for the broader purpose of venting his nation's pent-up rage, a complex of resentments that was generations in the making. It came out in a tremendous spew of invective. Mao's regime, this new government from an ancient land, wanted recognition, yearned to assume its place on the world stage. But it also seemed to want an apology.

As Wu later put it to an Australian journalist, "I had only felt that I was facing the imperialist state and its followers, and that they had invaded, bullied, and oppressed China for over a century. The memory of it was deep and bitter in my mind as well as in the minds of hundreds of millions of Chinese people. Now the Americans were deliberately scheming to subvert and annihilate New China. They had acted in such a reprehensible, such a vicious way, that there was no need for courtesy on our part."

If the Chinese delegation had ever offered a tantalizing last chance for peace, that chance was long gone. Wu's diatribe had only hardened American opinion. Appeasing Mao in any way, admonished *Time* magazine, would be akin to "kissing a buzz saw." The United States immediately froze Chinese assets and, a few days later, led an effort in the U.N. General Assembly to enact a resolution condemning Mao's aggression in Korea. In protest, Ambassador Wu stormed out of the hall. No mainland Chinese emissary would appear at the United Nations for more than twenty years.

Soon Wu and his fellow delegates were on a plane bound for Beijing, where they would be given a hero's welcome.

29

MORPHINE DREAMS

Toktong Pass

As the battle for Fox Hill smoked and chattered outside, Hector Cafferata, cooped up in the med tent, had become distraught. One of his buddies, a guy named James Iverson, was dying beside him. Iverson had been grievously wounded by a grenade that had torn into his rib cage. He'd lost far too much blood, and he was going fast. Iverson had told Cafferata, "Hey Moose, when I die I want you to take my boots." Cafferata was still in his stocking feet—he'd never recovered his boots from the hill.

A few hours later, Iverson passed. Cafferata slipped the boots from Iverson's feet. He tried to put them on, but they didn't fit—Cafferata's feet were "canal-boats," as he liked to say, size 14. So he got a knife and cut holes in the ends. Now, staring at his own frostbitten toes wiggling out the ends of Iverson's boots, Cafferata wept like a baby. A pair of guys came and collected Iverson's corpse. They threw it on the stack outside the tent.

As the hours drained away, Cafferata drifted in and out of consciousness. The morphine had a powerful effect on him. He kept having dreams, fevered and fluid and weird. Maybe they weren't really dreams, because he often wasn't asleep, but riding the ebb and flow of the injections. He dreamed about his father, who had been a paper engineer. Hector Cafferata Sr. had traveled the world building paper mills. Though he wasn't rich, he knew how to make the paper that money was printed on. He was of Italian origin, but he'd grown up in Argentina and Peru before immigrating to the United States. He'd taught Hector how to shoot, how to hunt. He had much kindness in him, but he could also be abusive—"a nasty little bastard," as Cafferata put it. Hector Sr. had grown up playing soccer and had a habit of kicking Hector when the boy crossed him, which was often. "I probably would have been a Rhodes scholar if I hadn't spent so much time taking his shoe outta my ass."

Sometimes Cafferata dreamed of football games. He'd be back on some gridiron in New Jersey, knocking people down, clobbering them. Though he wasn't particularly coordinated, he was so big and so strong that he intimidated the opposing players. But Big Hec didn't want to hurt anyone—he just loved the brawl and the bluster of the game.

He dreamed of the Chinese, too, especially the ones he had captured. He couldn't get those three kids out of his mind. He kept reliving the moment they first came to him, the moment they surrendered on the hill. He didn't see where they had come from. It was as if they had popped out of a hole in the ground. In the dream he kept having, he would hear Chinese voices, and then they would emerge from a thick cloud, like angels. He looked at them, studied them. Their faces were frozen, covered in snow. Icicles clung to their noses and chins. They were scared, probably freezing to death. They had their hands up, and they begged him to spare their lives.

But in the dream, Cafferata didn't take pity on them. In the dream, he swung his weapon around and shot them—*pow-pow-pow.* He didn't have to aim; he put a front sight on them and down they went.

The three Chinese soldiers kept reappearing in Cafferata's thoughts and reveries. "It was as though they were haunting me," he said. "It was like somebody was trying to tell me something."

≡

Captain Bill Barber was losing hope for Fox Company. The siege was unrelenting. The enemy understood how vulnerable the Marines were on this cheerless hill, how close they were to the end. The Chinese could nearly taste victory. Over the past few nights, Barber's company had been cut in half. He was down to 159 "effectives." He could hold out another night, he thought—maybe. After that, it was doubtful.

But Barber didn't show his despair. He was a stirring spectacle on the hill, hobbling around like a mad prophet, using a gnarled stick as a crutch. He did his damnedest to ignore the pain in his shattered pelvis. The corpsmen had treated the wound with sulfa powder, had bandaged it as best they could, then improvised a splint out of a pair of pine limbs. At one point he barged into the medical tent and was heard to say, "All right, men, here it is. Things are pretty bad. But I've seen 'em worse. We're not pulling off this hill unless we all go together."

That night, the Chinese brought out their amplifier once again, and a voice reverberated over Fox Hill. But this time it wasn't a Chinese voice—it was American. His name was Robert Messman, a lieutenant from an artillery unit based at Yudam-ni. He'd been captured a few days earlier. The jeep he'd been driving was found on the road a few miles south of Yudam-ni, without bullet holes or blood smears, no sign of a struggle. The Chinese had marched him into the hills to a tiny farmhouse. Now they were using Messman for propaganda purposes.

"Men of Fox Company, this is First Lieutenant Robert C. Messman of King Battery," the voice said. "I was captured two days ago by Chinese Communist Forces."

The Marines in their foxholes squinted at one another. Was this a hoax? No, they decided, the man was definitely an American—though it sounded like he was reading a script. "Men of Fox Company," the voice continued, "if you surrender now the Chinese will treat you according to the Geneva Convention. They will feed you. They will provide warm clothing. They will treat your wounds." Over and over, Messman urged the Marines of Fox Hill to give up the fight.

When his voice trailed off, the machine gunners shot off a few rounds, as if to tell the Chinese, in a language they would understand: *Fuck you.*

Captain Barber, meanwhile, had been stewing over what to do with his own prisoners. By this point, Fox Company had captured more than thirty Chinese soldiers, including the three frightened kids

Cafferata had collected on the hill. The prisoners were nearly freezing to death, they were hungry, they were lame with frostbite. Many of them had serious battle wounds. Some of them had already died of exposure.

The ones who were left sat on their haunches in a pitiful cluster, shielding one another from the wind. Barber could let them go, but who knew what would become of them? The Chinese officers might not take them back. Fox Company had seen commissars shooting Red deserters, and these prisoners—the uninjured ones, at least—might be perceived as such. But if their superiors did take them back, they'd put weapons in their hands and send them up the hill to kill more Marines.

Captain Barber had been avoiding the obvious, but he knew what he had to do. It was the hardest decision he would make in his entire life—harder than anything he'd been forced to do on Iwo Jima. He was a God-fearing man, a Christian, a churchgoer. But he thought he had no choice.

He called for a private from Georgia and told him to get a few other Marines and take care of the problem. And so they went around back, behind the command post and the med tent, back where the prisoners squatted in the snow. Then they shot every one of them in the head.

When Cafferata found out about it, he lost it. "Man, I was wild," he said. "I was hot. I told Barber, 'I'd be a son of a bitch if I'd shoot 'em. If you'd told me to do that, I would have given you my rifle.'"

30

NO SOFT OPTIONS

Hellfire Valley

Through the night of November 28 and into the next morning, General Smith's stronghold at Hagaru had tenuously held. His army of cooks and clerks and mechanics had done their best to repulse the Chinese attacks. But East Hill was now largely under enemy control, scores of Marines had been killed, and casualties were numbering in the hundreds. Worst of all, the airstrip project, which had only a few days to go before completion, was in jeopardy. General Song had well grasped the strategic significance of what the Marines were attempting to do at Hagaru, and he had dedicated more and more resources to overrunning this American bastion before it could become an American airport.

Smith understood that if Hagaru fell, the whole division could be wiped out. Smith had to have help, but he knew he could not look for it from his regiments to the north. Litzenberg and Murray were still tied down in Yudam-ni. And Litzenberg's Fox Company, at Toktong Pass, was facing destruction. Any relief would therefore have to come from the south, from the direction of Koto-ri. The unit based in Koto-ri was the First Regiment, under the command of a legendary officer. He was the most decorated and possibly the most famous Marine then alive. His name was Lewis Puller, better known as "Chesty."

Colonel Chesty Puller was a truculent little man, with a crabbed face, a ramrod physique, and an irrepressible spirit. He seemed to have been born curmudgeonly. He spoke in a thick Virginia Tidewater drawl. Reporters loved him, because nearly every word that came out of his mouth was quotable. He was famous for the blunt, off-the-cuff declarations he uttered in the field—"Pullerisms," they were called. In the Pacific, when he saw his first flamethrower, he was said to have barked, "Where do you put the bayonet?" He reportedly once exhorted, in the heat of battle: "This is a shitty war—but it's better than no war at all!" In Koto-ri, that very week, upon learning that a big airdrop of supplies, packed in Tokyo, contained condoms, Puller had been heard to growl, "What the hell do they think we're doing to these Chinese?" A few days earlier, when he learned that his regiment had become surrounded at Koto-ri, he had minted a new Pullerism: "So the Chinese are to our east. They're to our west. They're to our north. And to our south. Well, that simplifies things. They can't get away from us now!"

Puller's men loved him, and were deeply loyal. "The higher brass may not have appreciated him, but we sure did," said Manert Kennedy, a staff sergeant from Detroit. "When we were under assault, he'd be out there with us, walking the line. We'd get a tap on the shoulder: 'How ya doin', Marine? Hope you get to kill a lot of Chinese tonight.'" One day at Koto-ri, Puller was on a field phone, trying to learn the disposition of the enemy. "How many Chinese're up there on that ridge?" he yelled into the mouthpiece.

"A shitpot, sir," a scratchy Marine voice replied.

"Well," said Puller, "I'm sure as hell glad somebody up there can count!"

It could be said that Puller suffered from some of the same personality excesses that plagued Ned Almond. He was always aggressive, perhaps recklessly so, but that was a trait that seemed more appropriate for a regimental commander than for a general. As long as a superior officer kept a close eye on him, Puller could be a dynamic asset on any battlefield. He was as brave as he was optimistic. General Smith had a deep fondness for the man, although he viewed him a little skeptically, calling him a "picturesque" character.

Smith radioed Puller. What was the situation at Koto-ri? he asked him. How bad were things there? Koto-ri lay eleven miles away. Could he spare any men to come smashing up the road to reinforce Hagaru in its hour of need?

≡

The answer came back in the resounding way it usually did with Puller: *Hell, yes.* His First Regiment was under strain at Koto-ri, but he thought he could slap together a motley team of fighters to barge up the road. He could muster a force of maybe a thousand men. It would consist of some Marines and some Army guys, too. But the main fighting force Puller had in mind was a peculiarly staunch unit, 235 strong, from Great Britain. They were a group of Royal Marines known as 41 Independent Commando.

To the American Marines, these cousins from across the pond were a curiosity and a delight. They had arrived in North Korea in mid-November and had immediately been assigned to Smith's division. Trained for amphibious reconnaissance, they had an undeniable panache. They wore svelte green berets, a form of headgear that did nothing to dissuade bullets or guard against the cold. They actually said things like "Tally ho!" and "What's next on tap?" They kept up a spit-and-polish appearance and shaved almost every morning, a discipline the hirsute Marines found amazing, particularly in this cold weather. Their leader was a tall, stylish World War II veteran named Lieutenant Colonel Douglas Drysdale.

That morning, when Chesty Puller assigned Drysdale to lead a breakthrough force to Hagaru, he was game. Drysdale's Royal Marines were eager for a fight. The hodgepodge of Americans and Brits, dubbed Task Force Drysdale, jumped off from Koto-ri at 9:45 a.m. on November 29. Nine hundred men and a train of about 150 vehicles wound through the mountains, following the frozen Chang-jin River as it coursed along the valley floor. Drysdale called it "poisonous" country, perfect for ambush. He was right: Within the first mile, they ran into trouble. The Royal Marines and a company of American Marines—George Company, under the command of Captain Carl Sitter—took turns neutralizing Chinese positions in the heights while the rest of the column crept ahead. Drysdale, quickly ascertaining that he needed more firepower, radioed Puller, who sent some twenty tanks up the road. But even with the tanks in their midst, Drysdale found that his task force was barely inching forward—it was, said one Marine historian, "a joint-by-joint advance, like a caterpillar's."

The Chinese had cratered the road with bombs and had erected numerous barricades to make the way impassable. High along the hills, hidden behind Korean huts, the Chinese riddled the thin-skinned American vehicles. Their mortar shells walked up and down the convoy. In one tricky passage, a defile Drysdale had dubbed "Hell-fire Valley," the enemy assaults were especially withering. The men of 41 Commando absorbed the onslaught with regulation sangfroid, but by afternoon Drysdale had already taken dozens of casualties. With nightfall approaching, he assumed that the Chinese would only step up their attacks. He thought he should turn back—he sensed this was turning into a suicide mission—but he needed higher authority to do so. Drysdale radioed Hagaru to ask Smith if he wanted the endangered column to keep going or return to Koto-ri.

Smith understood Drysdale's dilemma, and he fully appreciated the difficulty of the relief mission he'd been assigned. But the general had been visiting Hagaru's hospital tents, had seen the rows upon rows of casualties from the previous evening's fighting. He believed Hagaru was in danger of falling that night. Now was the time for a bitter command decision that only he could make. He knew from his experiences on Peleliu and Okinawa that, at critical times, sacrifices were required of one group in order to save the whole. Later, when discussing how he'd weighed this decision, Smith said, "War leaves no soft options."

And so Smith's reply was radioed back to Drysdale: "Press on at all costs."

The British officer, though sobered by the command, didn't flinch. "Very well then," Drysdale said, "we'll give them a show."

And a show it was, a vicious fight every step of the way. "There were a lot of Chinese bastards shooting at us from every conceivable angle," said one Royal Marine. "The firing was coming in, thick and fast," said another. "Our lorries were totally exposed. The chap right next to me was killed, and there was a lot of messing about. When the going got tough, we just bashed on." The tanks in the front bulled through without much problem, with George Company and the leading edge of Drysdale's commandos following. Finally, sometime after midnight, they arrived at the margins of Hagaru, which was bathed in the glow of the floodlights the engineers were relying upon to build their airstrip. The town was engrossed in a major battle, but Drysdale, stumbling through the mayhem, managed to locate Smith's headquarters. At around one in the morning, he sauntered in. Blood dribbled from two shrapnel wounds in his arm. His green beret was fixed to his head at a rakish angle. Drysdale had no idea how many of the nine hundred men in his task force had survived, no idea how the rear of the long vehicular train might be faring. But he wanted Smith to know he was here. In the faltering light of a tent lantern, Drysdale gave a crisp salute and proudly announced, "41 Commando, present for duty."

By then, several hundred men had stumbled into Hagaru, and they'd arrived at a critical juncture. Almost immediately, Smith put the Royal Marines and Sitter's George Company to work. Smith pointed to the landmass that rose over the village, which now flashed with combat: East Hill. They were to do everything in their power to take it.

≡

The story of Task Force Drysdale would turn out to be a story of thirds. The front of the column, about four hundred men in all, would reach its destination—mauled, but still ready to fight. The rear third, encountering sharp resistance early on, had scrambled back to the safety of Chesty Puller's stronghold at Koto-ri. But the middle third of the column, trapped in the deepest folds of Hellfire Valley, encountered an altogether different situation. These men, some 320 in all, were doomed.

Somewhere in the thick of that middle section, a seventeen-year-old ammo carrier from Oklahoma named Jack Chapman was riding along in the late afternoon when a Chinese grenade plopped into his truck bed. It clunked on the floorboards, right between his legs. He stared at it wide-eyed. Lucky for everyone, Private Chapman was young and had quick reflexes. He grabbed the grenade without thinking and flung it over the side of the truck. It landed in a ditch and immediately detonated; no one was hurt. Everyone thanked the Okie for saving their lives. But this proved to be only the beginning of a long night of horrors.

Chapman was a scrappy kid, with a bony face and a sweet, snaggle-toothed grin. Part Cherokee Indian, he'd grown up in poverty. When he started butting heads with his stepfather, he was passed around to various relatives. He dropped out of high school and became more or less a vagrant, taking odd jobs around the country: gathering walnuts and pecans, picking cotton, working his uncle's onion fields in Michigan, setting pins at a bowling alley in Tulsa. He hated this work, hated moving around all the time. He knew his life needed direction. So he lied about his age—he was sixteen at that point—and joined the U.S. Army. He trained at Camp Chaffee, Arkansas, then at Camp Carson, Colorado. His outfit spent a couple of months in Alaska before transferring to Wisconsin.

Chapman got orders to report for duty in Japan, then was sent to Korea with the Seventh Infantry Division's Thirty-first Regiment. The Army had first sent him north, and his unit briefly lingered at the banks of the Yalu, but then he was sent back to Hamhung. Through convoluted routings, Chapman and his comrades happened to be in Koto-ri the day Task Force Drysdale was mustered. It was a perfectly random circumstance that he'd fallen in with a group of British commandos and American Marines.

Now he was in Hellfire Valley, a quick-witted teenager wiser than his years. Shortly after the sun went down and the road was cloaked in darkness, the column slammed to an abrupt and eerie halt. Then Chapman's truck came under fierce attack—apparently the whole convoy did. Leaning out, he could see the main problem: A truck up ahead had been hit by a mortar and was engulfed in flames. The wreck wasn't budging, and because this was a single-lane road, an immovable logjam had been created. The Chinese knew what they were doing; they were trying to fractionalize the column, to snip it into pieces so that they could then attack each isolated fragment with impunity.

Everyone in Chapman's truck jumped out and dropped into a

ditch. They grabbed their rifles, and a battle began. Chapman could see hundreds of Chinese pressing in. Bullets stuttered across the snow. Chapman had a carbine, but the little rifle wasn't reliable in the cold. He might as well have been plinking away with a BB gun.

In his immediate vicinity, only one instrument of any significant firepower seemed to be operating. It was an M20 75-millimeter "recoilless rifle" mounted on the back of a truck. To call this weapon a rifle seemed a bit of a misnomer: It was a formidable piece of tube artillery that could accurately project a twenty-pound shell a distance of more than two miles. Upon detonation, the long, slender, rocket-shaped missile broke up into innumerable hot fragments, visiting extreme punishment on anyone within its range. Rather than having a recoil mechanism, like most artillery pieces, the M20 relied upon explosive gases to propel the shell forward, so the weapon had almost no "kick" when it was fired. Instead, a miasma of noxious gas shot out of the back of the gun, a plume long and hot enough to scorch anyone who stood in the way. It was a wicked weapon, and one that Chapman happened to know how to use.

As the fighting intensified, the Army sergeant manning the M20 began to lose his nerve. He was a perfect target up there, blasting away with this bright, powerful weapon, splattering shrapnel over the hillsides. Wielding that gun was an advertisement to be shot; surely the Chinese would take their vengeance soon enough. Finally the sergeant could take the pressure no more. He jumped down from the truck and knelt in the snow, then started to pray out loud, intoning in a strange voice, "Protect us, O Lord. Protect us, O Lord."

Captain Peckham, the commanding officer of Baker Company, scolded the sergeant, cursed him, shamed him. "Get back up there, you coward!" he yelled. But the sergeant wouldn't budge from the ground, even when threatened with court-martial. So Captain Peckham stomped among the men, looking for a volunteer. "Anyone know how to work this gun?" he demanded.

Haltingly, Jack Chapman raised his hand. Back in Colorado, he had trained on the M20. He thought he could remember how to use it. He was the only one brave enough, or dumb enough, to give it a try.

≡

Chapman crawled up on the truck bed, leaving himself exposed in the moonlight. He swiveled the big gun and aimed at the distant muzzle flashes. The first shell shot through the frigid air and smashed into a ridge several hundred yards off, spitting its destruction in a wide arc, well behind enemy positions. He realized he would have to use the weapon close-in—much closer than it was intended for. Chapman aimed at a group of enemy soldiers who came bursting across the slopes, thirty or forty yards away. He fired, and the projectile cut them all down.

Manning a weapon as lethal as this, he knew it would be only a matter of time before the Chinese found him. First they shot him in the left arm, then the right leg, then the other arm. Another bullet, he later learned, had lodged in a pack of Philip Morris C-ration cigarettes stuffed in the breast pocket of his field jacket. A British medic quickly treated his wounds, and Chapman climbed back on the truck. He was shot again, in the left hip, and then he took several slivers of shrapnel. He was bleeding profusely—he wondered how much blood a person could lose before he died. But his adrenaline was pumping, and somehow it blunted the pain. He kept at his gun, blasting into the night.

He had finished reloading the recoilless rifle when he sustained his seventh—and final—wound. As he prepared to fire, a bullet from a burp gun struck his forehead. It knocked him off the truck and rendered him unconscious. He lay sprawled in the snow. The bullet had embedded itself in his cranium, but, amazingly, it hadn't entered his brain.

When he came to, the fighting was nearly over. The Chinese had overrun his unit's position, and now they were looting the trucks, rummaging for food and ammunition. Chapman looked up at one point and saw several enemy soldiers studying the recoilless rifle. They fussed with it, aiming it this way and that; they were trying to figure out how the thing worked. Several of them were standing behind the gun when it accidentally went off. The backblast of propellant gas scorched them horribly.

It was a grisly sight, and yet Chapman, dazed and bleeding on the snow, had to stifle the urge to smirk. But the battle of Hellfire Valley was over. What was left of Task Force Drysdale had been severed into various surrendering groups trapped on the road, each one having lost contact with the others. "Press on at all costs," Smith had said, and the costs had been steep. More than a hundred American and British troops had been killed, with 150 wounded. Seventy-five vehicles had been wrecked—many of them were crackling in flames across the valley. More than a hundred men were laying down their arms.

The highest-ranking American in the contingent, Marine major John McLaughlin, was negotiating terms for surrender. He stood on the roadside with three Chinese officers, one of whom spoke fluent English. Trying to stall for time, McLaughlin, rather ludicrously, told the Chinese man that he was now willing to accept *their* surrender.

The Chinese were not amused. "You have five minutes," the English speaker said, holding out his hand and counting off his fingers. "If you do not surrender, we will kill you all."

By this point Jack Chapman had passed out, and the next thing he knew, he and ten other prisoners were hunched on the dirt floor of a tiny Korean farmhouse somewhere on a hillside. Their bodies reeked in the close heat, and Chapman's numerous wounds, having thawed, began to flow again. He wasn't sure how long he remained in the farmhouse. Drifting in and out of consciousness, he lost track of time. But at some point, the Chinese prodded him and the other men to their feet and herded them out of the structure. They joined a larger group of captives, both British and American, and marched for nineteen days, following obscure mountain trails until they reached a barbed-wire enclosure at a place called Kanggye, not far from the Manchurian border. Chapman would be a prisoner of war for nearly three years.

31

ONE-MAN ARMY

Yudam-ni

The eight thousand Marines at Yudam-ni had held on handsomely since the first attacks on the night of the twenty-seventh, and now they were planning their breakout for Hagaru. Colonel Homer Litzenberg, the Seventh Regiment commander, and Lieutenant Colonel Ray Murray, the Fifth Regiment commander, had been working closely on the plans, and they were just about ready. They aimed to break out the following day, December 2, and bash their way toward General Smith.

But Litzenberg had a more immediate concern to consider. He was deeply worried about Fox Company. After four days of attacks at Toktong Pass, Fox was on the verge of annihilation, and the colonel knew it. He had been on the radio with Barber that morning, with a scratchy signal, and although the wounded captain had put his best face on things, Litzenberg sensed that Fox could not hold on for another twenty-four hours. Three-quarters of Barber's company had become casualties. His men were shell-shocked. His perimeter had shrunk, and shrunk again, like a hardening tumor. What Barber faced was a concentrated version of the same situation that was playing out all across the Chosin battlefield. The Chinese bodies continued to pile up on the slope, yet the enemy, sensing victory, pressed with renewed vigor. Fox Hill, Litzenberg feared, was about to become the site of a massacre.

He knew something drastic had to be done to save these surrounded men. Around midday on December 1, Litzenberg summoned one of his most experienced and resourceful battalion commanders,

Lieutenant Colonel Raymond Davis, to discuss the predicament. Ray Davis hailed from Georgia and was a taciturn man, cool under pressure, with "eyes that bored into you," according to one subordinate. He was a war hero from the Pacific who had fought admirably at Guadalcanal, New Britain, and finally Peleliu, where he won the Navy Cross. He had seen harrowing situations, but none so bad as the one Fox Company faced. Sitting on a rickety cot in the glow of a ticking stove, Davis and Litzenberg squinted at a topographical map. "The Chinese think we're roadbound," Litzenberg said. "They think we'll stick with our vehicles."

Davis understood that he was being asked to concoct a plan. "What if we get *off* the road and take the high ground?" he said. "What if we sneak up from behind?" He was already anticipating a scheme that had been taking shape in Litzenberg's mind all morning: The colonel wanted to commit Davis's First Battalion, as many as five hundred men, to save Fox Company. He wanted Davis to lead them off the road and march them, at night, through trackless mountains to relieve the Fox defenders.

The idea stirred Davis's blood. It was the kind of assignment that would rally any Marine: a stealth mission in enemy territory to rescue comrades in trouble. Davis studied the map a little more closely. A straight off-road journey from Yudam-ni to the back door of Toktong Pass looked to be only five miles as the crow flies. But this wasn't going to be a direct shot. It appeared the trek would involve trudging over at least three steep ridges. With the detouring that such variegated terrain would require, Davis thought it would likely be a march of ten miles—maybe more. Who knew if a column of men already so frostbitten and fatigued could make such a march? Who knew if this sketchy map bore the scantest resemblance to the country? Who knew how many thousands of Chinese troops were encamped on those intervening ridges? This could be a mad mission, Davis thought. In trying to save a company, a battalion could be wiped out.

Litzenberg understood the risks but demanded that they see the idea through. Idly presiding over the slaughter of Fox Company was not an option; it would weigh on his conscience for the rest of his life.

Another factor further cried out for an overland march: Litzenberg understood that Fox Company was no longer in full control of Toktong Pass. The next day, when the nearly eight thousand Marines made their break from Yudam-ni to Hagaru, with all their wounded and their vehicles, they would have to slip through this narrow place. It was impossible to bypass. The Chinese, holding the high ground on both sides of the MSR, could inflict terrible damage on these road-bound Marines. The CCF might even succeed in halting the column—which, within the confines of this highland choke point, could result in a bloodbath. But if Davis could make a cross-country march for Fox Hill and swiftly seize the pass, much carnage could be avoided. Davis's First Battalion would, in effect, unlock the back gate—and keep it open for the main body of withdrawing Marines.

Litzenberg wanted Davis to work up a plan and bring it back as soon as he could. Davis was determined to try. His reasoning carried a powerful simplicity. "Some fellow Marines were in trouble," he later wrote. "We were going to rescue them and nothing was going to stand in our way."

≡

It didn't take Davis long. Within a half hour, he reported back to Litzenberg, who gave his brusque assent. Davis gathered his company commanders to discuss the details. It would be a "bold dash" under the stars, Davis said, and they would have to "streamline the battalion." Surprise, he said, "will be our essential weapon. Marines don't ordinarily attack at night, so the Chinese won't be expecting us." Davis wanted them to veer off the road just south of Yudam-ni and climb into the highlands, moving out in single file, heading southeast under cover of darkness. Utter silence was imperative. There could be no rattling of weapons along the route—everything would need to be muffled in cloth and tied tight. No cook fires would be allowed on breaks; the men were to bring only quick-energy food, ready to eat.

They'd travel as lean as possible, carrying mostly light arms, with only two 81-millimeter mortars and six heavy machine guns in the whole column. They would jettison nonessential equipment in favor of extra ammo. Each Marine would tote an 81-millimeter mortar shell in his sleeping bag. More ammunition could be hauled on stretchers. Most Marines were expected to lug an additional bandolier of machine gun bullets.

Communication would have to be kept to a minimum. The marchers would whisper orders down the column of men—no yelling, and no radios unless absolutely necessary. Davis predicted casualties, and he'd decided to bring along an excellent regimental surgeon, Navy lieutenant Peter Arioli, to treat them. The dead were to be buried in the snow where they fell, with the wounded placed on stretchers as the ammo originally hauled in them was expended.

To help with navigation over the confusing array of ridges, Davis devised an ingenious solution: Every so often, a howitzer in Yudam-ni would lob a star shell toward the southeast, along a predetermined azimuth, a line leading straight to Fox Hill. In theory, all Davis's Marines had to do was follow the intermittent bursts of white phosphorus as they arced across the sky. The marchers would be guided to their destination like wise men to Bethlehem.

One hitch presented itself: Davis's men would likely have to fight a battle even before they could leave the road. By the look of the map, the best spot to jump off into the mountains was a place called Turkey Hill. The Chinese were known to be dug in there, and it would require blunt force to bust through their position and "destroy" them, as Davis phrased it. But he was confident that it could be accomplished by dusk. Then they'd be on their way, threading through the mountains toward Fox.

But Davis had one more important decision to make. Who would lead the march? Which platoon? Davis had decided he would place himself near the middle of the column, making sure it held shape, sometimes floating forward, sometimes floating back toward the rear to exhort the stragglers. The line of marchers would stretch over a half mile. In the darkness, with the view ahead obscured by so many knobs and ridges, which themselves might be obscured by patches of ice fog or slanting snow, each Marine would have no choice but to follow the dim form immediately in front of him. The march's success would crucially depend on who was in the vanguard, choosing the path forward.

Davis didn't have to think very hard about who it would be. He would pick the most irascible and most determined officer in the whole battalion, the twenty-four-year-old leader of Baker Company's Second Platoon—a man who, coincidentally, spoke Chinese. His name was Lieutenant Lee.

≡

As a boy growing up in Sacramento, the firstborn son of Cantonese immigrants, Chew-Een Lee had loved his father's collection of ancient Chinese history books. He was especially fond of the turbulent period known as the Three Kingdoms. He feasted on the epics from that era, which recounted fabled generals, misty battles, and miraculous feats of military strategy. Lee was equally impressed by the tattoo that his father wore on his arm—DEATH BEFORE DISHONOR, it said. As a kid, he didn't know exactly what it meant, but the words imprinted a stark ethos upon Lee's imagination that followed him through life. He knew to the point of certainty that he would die a heroic death for his country on a battlefield far from home. It was the central assumption that propelled him.

Lee was said to be the first Chinese American officer in the history of the United States Marine Corps. People didn't appreciate his toughness at first glance. He was a slight man, five foot six, fine-boned, angular of features. He had a birdlike quality: a scowling face; a sharp beak of a nose; long, bony fingers; and a certain jerky precision in his motions. His disapproving eyes, tight and assessing, seemed to catch everything. His sense of humor, said one historian, "was not well developed." "He was hardly a congenial companion," said a mortarman who served under Lee. "But on duty he had our full respect as an unequivocal professional."

Lee had joined the Marine Corps in the waning days of World War II, but instead of being sent off to fight, he was dispatched to language school to become a translator. It chafed him to no end that the Marines refused to place him in a battlefield role—he viewed it as a blot on his honor. From then on, Lee had a chip on his shoulder. As he rose through the Marine ranks after the war, he only grew in his resolve to prove that he was the best, the most scrupulous, the most aggressive, the most eager to take action. "Some say I was by the book," he said, "but I tell you, it *works*. Fire and maneuver! Discipline! Esprit!"

What Lee wanted most—what he demanded—was respect. He was not some interpreter, not some coolie with an unctuous smile curling across his face. He was determined to lead a unit of American men into combat, to have a group of Marines "willingly follow me into hell," as he later put it. Sinophobia, dating back to the days of the "Yellow Peril" and the Chinese Exclusion Act, ran deep within the United States, the country to which Lee's merchant father had moved as a young man from Guangdong Province in the 1920s. Lee himself had grown up in Depression-era California as a hybridized citizen, steeped in the old ways but also those of his native land. Whatever barriers Lee had broken in the Marine Corps he hadn't broken bashfully. He was a defiant man, and very much old-school—as old-school as those martial epics from the Three Kingdoms.

The Korean conflict, Lee felt, gave him a second chance to prove himself in war. If some Marines wondered whether he could be trusted fighting soldiers from the land of his heritage, those worries were quickly expunged on the battlefield. On November 2, during an engagement at Sudong Gorge, Lee had become the protagonist in a spectacularly heroic exploit. Attempting to trick the enemy into revealing his position high on a wooded hill, Lee launched a one-man assault in which he tried, in effect, to impersonate an entire platoon. He would fire a round from his carbine, then scurry to another position and hurl a grenade, then run to still another spot and shoot his pistol. It was a ruse. He was trying to produce a lot of racket over a broad area in order to make the Chinese think a large group of Marines was advancing.

The stratagem worked: The enemy returned fire. Lee, now knowing where the Chinese bunker was hidden, crawled right to it. When he drew within a few yards and realized he'd been detected, he blurted out, in Mandarin, "Don't shoot—I'm not the enemy!" The distraction bought him enough time to hurl his last two grenades. Then he opened up with his carbine, at full automatic, spraying the

Chinese. In a few moments, it was over. Single-handedly, First Lieutenant Lee had captured the enemy outpost and secured the hill. He was a one-man army. For this action he would win the Navy Cross.

$$\equiv$$

The next morning, November 3, while still fighting near Sudong Gorge, a Chinese sniper shot him. The bullet struck Lee's right elbow and spun him around. He crumpled to the ground in agony. His arm had a gash four inches long, with the shattered bone exposed. A corpsman bandaged him and wrapped his crippled appendage in a sling, and he was sped, against his wishes, to an Army evacuation hospital, set up in a school gymnasium in Hamhung, to await transport to Japan. But after a few days languishing in this makeshift infirmary, Lee became restless and bored. His arm may have been wrecked, but he could still fight—he was sure of it.

The day before he was to fly to Japan, Lee and another convalescing Marine hatched a plan to escape from the Hamhung hospital and return to the front lines. Finding a supply of weapons piled in a nearby compound, they armed themselves and took off in the first jeep they could scavenge from the motor pool. "Hey, that's mine!" a soldier yelled as they pulled away. Lee waved his carbine in the man's direction, and they disappeared into the backstreets of Hamhung, laughing like two fugitives as they careered toward the outskirts of town. They drove along a levy between dry rice paddies and aimed north, for the snow-dusted mountains.

When the jeep coughed out of gas, they got out and walked the last miles to the front lines. Lee found his unit, Baker Company, and reported for duty, his slung arm encased in a heavy cast. He was worried he might be court-martialed, not only for violating doctor's orders and going AWOL from the infirmary but also for commandeering—stealing?—an Army vehicle. But Baker Company welcomed him back and said the paperwork would be taken care of.

Lee was thrilled. "I wasn't yet convinced," he said, "that I had earned any glory." He wasn't afraid to die in combat—he still believed it was his destiny, now more than ever. "I was prepared to meet my

Maker," he said. "I could care less about heaven or hell. If I were to die, it would be a pulling of the blinds, and then there would be darkness."

Lee had worn his sling and his cumbersome cast through the fighting at Yudam-ni. He had led numerous patrols, having to hold his weapon with his left arm, steadying it with his hip. He had never complained. Lee, said one noncom who served under him, "was hard as steel, tough as nails, cold as ice, and reliable as time itself—all the clichés apply."

When he learned that Colonel Davis had picked him to lead a night march through the mountains to rescue the men of Fox Company, Lee wasn't surprised. On the contrary, he expected it. "This was almost a mission impossible, over unknown terrain, against unknown forces," Lee later said. Davis's choosing him, Lee thought, seemed "sort of logical." Not because he was Chinese American, not because his language skills might come in handy, but because he was the best.

32

EVERY WEAPON THAT WE HAVE

Washington, D.C.

On the morning of November 30, the Indian Treaty Room inside the Old Executive Office Building was crowded with more than two hundred journalists. President Truman marched into the great room and, in a blaze of flashbulbs, assumed his place at the dais. Then, in a crisp, straight-ahead manner, he proceeded to read a prepared statement. "Recent developments in Korea" he said, "confront the world with a serious crisis." China, he noted, had mounted a strong and well-organized attack against the United Nations forces, "despite prolonged and earnest efforts to bring home to the Communist leaders of China the plain fact that neither the United Nations nor the United States has any aggressive intentions toward China." The "historic friendship" between the United States and China made it all the more shocking to Truman that Mao had sent his armies to fight against American troops. The prospect for negotiations did not seem promising, Truman said—the representatives of Communist China had given no indications that they were willing to talk. Given this, the United States had no choice but to prepare for full-scale war. "It is more necessary than ever before for us to increase at a very rapid rate the combined military strength of the free nations," Truman insisted. He would call for dramatically expanded funding for all branches of the armed forces and would request a substantial increase in the budget for the Atomic Energy Commission. Truman concluded his statement, saying, "This country is the keystone of the hopes of mankind for peace and justice. We must show that we are guided by a common purpose and a common faith."

Then the president opened the floor to questions.

Q. Mr. President, in what detail were you informed about these [latest developments]?
THE PRESIDENT. Every detail. On November 23 General MacArthur had launched an assault on the Communist forces in Korea in an attempt to end the war. On November 28 he issued a special communiqué stating that the United Nations forces faced an "entirely new war" with an enemy force of 200,000 men.

Q. Mr. President, there has been some criticism of General MacArthur in the European press—
THE PRESIDENT. Some in the American press, too, if I'm not mistaken. They are always for a man when he is winning, but when he is in a little trouble, they all jump on him . . . He has done a good job, and he is continuing to do a good job. Go ahead with your question.

Q. The particular criticism is that he exceeded his authority and went beyond the point he was supposed to go.
THE PRESIDENT. He did nothing of the kind.

Q. Mr. President, since the Chinese delegation has shown no inclination to resolve the difficulties, what can be done then?
THE PRESIDENT. We are still working on the thing from every angle. The best thing that can be done is to increase our defenses to a point where we can talk—as we should always talk—with authority.

Q. Mr. President, will the United Nations troops be allowed to bomb across the Manchurian border?
THE PRESIDENT. I can't answer that question this morning.

Q. Mr. President, will attacks in Manchuria depend on action in the United Nations?

THE PRESIDENT. Yes, entirely. . . . We will take whatever steps are necessary to meet the military situation, just as we always have.

Q. *Will that include the atomic bomb?*
THE PRESIDENT. That includes every weapon that we have.

Q. *Does that mean that there is active consideration of the use of the atomic bomb?*
THE PRESIDENT. There has always been active consideration of its use. I don't want to see it used. It is a terrible weapon, and it should not be used on innocent men, women, and children who have nothing whatever to do with this military aggression. That happens when it is used.

Q. *Mr. President, I wonder if we could retrace that reference to the atom bomb? Did we understand you clearly that the use of the atomic bomb is under active consideration?*
THE PRESIDENT. Always has been. It is one of our weapons. . . . We have exerted every effort possible to prevent a third world war. . . . We are still trying to prevent that war from happening.

33

THE RIDGERUNNERS

In the Mountains South of Yudam-ni

Slightly before nine o'clock on the evening of December 1, Lieutenant Colonel Ray Davis moved among his men, reminding them of the significance of the mission upon which they were about to embark. "Fox Company is just over those ridges," he said, pointing toward the south. "They're surrounded and need our help."

Davis pivoted and gave a nod to Lieutenant Chew-Een Lee. Lee raised his good arm into the frozen air so everyone could see and cried out into the squalling night. Then he started marching, and the men of the First Battalion followed him into the hyperborean bush—450 Marines, in single file, groaning under their burdens like so many mountain yaks as they stomped through virgin snow. First came Baker Company, then Davis's command group, then Able Company, then Charlie, then How.

It was rough going from the outset, yet the men found a certain satisfaction, a sense of pent-up release, in the mere act of movement. For the past four days they'd been holding on at Yudam-ni, waiting and watching and fending off the enemy waves. Now they were attacking, a martial mindset that Marines tend to find much more to their liking.

Their pathfinder, their focal point, was hard to miss: Chew-Een Lee had chosen to wear a weird, garish outfit that made him look like a court jester. He had procured two panels that the Marines used for marking ground targets during air supply drops. These strips, designed to be spotted from the sky, were colored electric pink. Lee fashioned them into an ungainly harlequin suit that screamed, "Shoot me!" The getup looked ridiculous and also made him vulnerable to attack, but his devotion to practical purpose overrode any sense of fashion or fear. "I wanted my men to be able to refer instantly to me," Lee said. "I wanted to show my men that I wasn't afraid of enemy fire—and that they shouldn't be, either."

Apart from this, Lee also wore his clunky cast, still in a sling. He was in enormous pain, the unhealed bones of his right arm grinding with each footstep. Though Lee weighed only 120 pounds, he carried nearly eighty pounds in gear—including his various weapons and four grenades. He held a contour map and a compass in his left hand and picked his way forward, accompanied by three scouts arranged in a diamond pattern. It was especially exhausting for him and the others out front—they weren't just hiking; they were breaking trail, floundering in drifts, punching through snow that was often knee-deep.

Those in the middle section of the line found the trail nicely packed, while the men toward the rear encountered an icy chute, slick as a toboggan run. They scrabbled and slipped, sometimes reaching for the man in front and yanking him to the ground, too. In treacherous places, where the grade steepened to forty degrees or more, the men had to crouch on hands and knees and claw their way up, fumbling in the snow for anything to hold on to—stumps, roots, rocks, shrubs. They worked up a sweat while ascending the steeps, but when they headed down the rearward slope, their accumulated perspiration tended to freeze. Then they lurched like mummies, their ice-stiffened clothes crunching with every stride.

Yet they kept moving, animated by a single thought: If they didn't reach Toktong Pass by morning, there might not be a Fox Company left to rescue. By going overland, by divorcing themselves from the road that had dictated their movements, they started to see themselves in a different light. They were taking a page from the Chinese battle plan—abandoning their machines and vehicles, moving at night, blending with the high country. They acquired a new nickname, too. From that night on, the men of Davis's First Battalion became known as the Ridgerunners.

≡

Every five minutes or so, a star shell came sizzling overhead, lobbed from Yudam-ni. The shell would tint the slopes with a phantasmal glow as it streaked across the bowl of the sky. As the men headed into the mountains, they took comfort in knowing that they were still tethered to the regiment; comrades four miles away were thinking of them, lighting their way forward. Whenever Lee was high on a ridge, with an expansive view, he found the star shells helpful, but when he was down in a ravine, his vision closed in, the phosphorus rounds were nearly useless. With the mountain obscuring their trajectories, he could only consult his compass and guess his way forward.

Still, the Ridgerunners kept slogging ahead, deeper into the night. They were stretched out over the mountain now, too far apart to allow communication from front to back, or vice versa, should the need arise. Like a trail of ants, their single job was to trust and to follow. As one Marine later wrote, they formed a "long wavering line of dark lumps creeping up this ridge, down into the next valley—sliding, staggering, digging their heels into the snow to break their momentum." They slipped and fell, stood and fell again, "raising a chilling clatter in the darkness with helmets clanging against rifle muzzles."

In the lulls between gusting winds, an eerie silence fell over the high meadows. The only sound was the scuffing of boots through snow and the steady rustling of gear. The men lost themselves in their marching—some felt they'd slipped into a dream state. It seemed as though these mountain fastnesses had never been visited by man. They were uninhabited, perhaps uninhabitable. This world was made for deer and fox, for the Siberian tigers that preyed on them, but not for people. Sometimes the trees would explode from the severe cold, their trunks emitting a concussive snap. Other noises carried with stentorian force over long distances. A cough, said one Marine, sounded like a mortar shell.

The hours themselves seemed to congeal. Addled by the cold, stupefied by the monochromatic wastes of ice and rock, the battalion progressed in a blur. "Time had no meaning for us," wrote Joe Owen, of Baker Company. "We labored through infinite darkness in ghostly clouds of snow over an icy path that rose and fell but seemed to lead nowhere. We saw only the back of the man ahead, a hunched figure in a long, shapeless parka, whose every tortured step was an act of will."

The night sky constantly rearranged itself. Scarves of clouds and fog whirled overhead, and in the gaps, a bright moon seeped through. The mountain air danced with tiny ice crystals that shimmered in the moonlight—diamond dust, the phenomenon was called. On the ridges, the winds were especially keen, biting into the Marines, curing their faces, turning their lips raw. The temperature had dropped to twenty-five below zero, but with the windchill, it was later estimated at seventy below.

When the winds died down, Lee could hear the voices of enemy soldiers in the brittle air. Then he could see them—or at least the crests of their fur caps, peeking out of little holes. They looked like prairie dogs, he thought, nervously popping their heads up and down. They were no more than twenty-five yards away, yet they couldn't see the Marines. Lee could make out their alarmed conversation.

"Ching du ma?" (Do you hear something?)

"Tara da?" (Should we attack?)

At this early stage of the march, Lee sought to avoid a firefight at all costs. He skirted the Chinese and led the column out of harm's way. But hearing them talking had quickened the Marines' blood and pulled them out of their half-frozen apathy. "The voices had an adrenaline effect on us," said Owen. "Now we were charged with energy, our minds cleared."

≡

It was imperceptible at first, but as the march unfolded, Lieutenant Lee started to veer off course. Some tiny miscalculation or series of miscalculations must have compounded through the night. Maybe, in part, it was because Lee's compass was misbehaving, the needle floating erratically in its housing. He wondered if the cold temperatures were affecting the instrument, or perhaps a nearby vein of iron ore, which had once been mined from the surrounding mountains, was distorting the magnetic field. Lee finally decided that the compass was responding to the heavy concentrations of metal—weapons, mortar shells, ammunition—the Ridgerunners were carrying. Whatever the case, he couldn't get a good reading. He seemed to be slipping off toward the west, dropping downslope in the direction of the road.

Davis, who was marching nearly a half mile back and had a better perspective on the column's direction, could see this trend taking shape. He grew alarmed: Lee was slowly swerving toward a no-man's-land that, Davis happened to know, was scheduled to be bombarded with heavy artillery from Yudam-ni later that night. If Lee didn't correct his course, the front of the column could be ripped apart by friendly fire. His mistake, however slight, was imperiling the mission.

Davis tried to convey the message to Lee by radio, but the cold had already sapped the batteries. Then Davis tried to send word up the line, to be whispered from man to man to man. But in the lashing wind, with everyone so snugly bundled they could hardly hear a thing, this proved impractical. What began as a clear and urgent message degraded into a babble of caveman grunts.

When the westward drift became more pronounced, Davis decided to leapfrog ahead in hopes of reaching Lieutenant Lee. He kept bumping into his men as he marched past the long line, knocking some of them down in his haste. They cursed at this rude and impatient man, not knowing he was their commanding officer, bent on completing a crucial errand.

An hour later, Davis, gasping for breath, finally overtook Lee. The lieutenant was instantly recognizable, even through the snow flurries, by his harlequin suit of bright-pink panels. At first, the exhausted colonel forgot why he had come. The two men stood in the cold, bewildered, regarding each other like a pair of drunks. Finally, Davis was able to summon the message from the frozen banks of his memory: The march was dangerously off course. They had to readjust, toward the east. The howitzers from Yudam-ni would soon be raining fire upon this slope.

Lee tried to process this. He halted his scouts while Davis dropped into a nearby pit that had been gouged from the ground, apparently by Chinese soldiers. There, Davis covered himself with a poncho to escape the roiling snow. He snapped on a flashlight and studied his smudgy Japanese-issue contour map, which, at a scale of 1:50,000, could not have offered much help. But somehow Davis decided on a new course—"He must have had a crystal ball down in there," Lieutenant Lee marveled.

Davis confidently stood up to issue his new orders. But in those few seconds, he had already forgotten what he'd decided while deliberating in the hole—his brain was addled by the cold. So Davis returned to the pit, draped himself with the poncho again, and started over. A few minutes later, he stood up once more. This time having retained his fragile tissue of thought, he proclaimed the new route. He instructed Lee and the other officers to repeat what he'd said, because now that he'd uttered it, the information had escaped his recall.

Davis's mind seemed to be working at one-eighth speed. The neurological circuitry wouldn't fire. As Davis later phrased it, "It was too cold for government work." Definitely too cold for a southerner like himself, Davis thought. He wondered whether, had he hailed from Minnesota instead of Georgia, it would have made a difference.

Lee grudgingly accepted that his line of march had perhaps drifted on account of his fitful compass. But he was too proud to admit anything more. He insisted that he wasn't lost—and that he never had been.

A platoon of Marines, waiting for the new orders from Davis, had been resting in the snow nearby. The cold had descended upon them, and now they were virtually in a coma. They looked, said one account, like "frozen Buddhas" and seemed to have slipped into "that seductive, mindless mist" that accompanies advanced hypothermia. Davis, genuinely fearing that he might lose them, marched around the perimeter and slapped the prostrate Marines to get their blood going again. He tried to shake them into a state of alertness and haul them to their feet.

"What unit are you in?" he barked at one young man, but the kid looked at him blankly. He didn't know the answer.

The Chinese, too, were succumbing to the murderous cold. Not far away, a dozen hummocks studded the snow. When one of them seemed to twitch, a Marine sergeant bolted over to investigate. He dusted off the snow and found that it was a Chinese soldier crammed into a foxhole. The sergeant grabbed the man by the scruff of his neck and snatched him out. He was nearly dead. His eyes moved in their sockets but registered nothing. He wore sneakers but no socks. The sergeant examined the other humps and discovered that they were all Chinese soldiers, frozen stiff. The Marines could only look on them with pity. They had wedged themselves into their holes and died. The men of Davis's battalion knew that if they weren't careful, the same fate would await them.

It was nearly midnight. With the new orientation fixed in his mind, and on his map, Lieutenant Lee again took his place at the point of the column and resumed the march, aiming for a cathedral of rock that rose through the dervishes of snow.

=

At the top of the next ridge, Lieutenant Lee debouched into a snowy meadow that was strewn with boulders and fingers of granite. He and his point team tracked warily through this labyrinth of rock, for it seemed a perfect place for an ambush. Sure enough, a few minutes later, the Chinese opened up. The hills above flickered with muzzle flashes. Bullets spanged off the rocks and kicked up snow but did little damage. The Chinese must have thought this was merely a Marine platoon out on a midnight patrol; they had no idea that an entire battalion was coiling into their midst.

Lee signaled his squads to form into skirmish lines. They got on their knees and crawled up the slope. Lee and his men reached the top and poured across the enemy position, catching the Chinese off guard. Many of them were in their sleeping bags. The Marines ran among them, bayoneting them, smashing them with rifle butts, clubbing them with entrenching tools, firing at point-blank range. Some of the Chinese got away, dodging bullets as they darted among the boulders. Others hid in crevices and alcoves.

The Chinese who remained fought bravely—some hurled stones; one of them wielded a large tree limb—but they were no match for the well-armed Marines. Wrote Joe Owen: "The night was against them. We had the advantage of surprise and momentum. We fought with fierce energy, now released from the hours of cold and misery. The Chinese could do little more than try to escape."

Lee, as usual, was at the center of the fray. Encountering three Chinese soldiers ten feet away, he held his carbine with his left arm and fired. Two fell dead, but the third enemy soldier leveled his weapon at Lee, who must have presented a tantalizing target in his Day-Glo vest. A Marine sergeant standing nearby perceived the threat just in time and dropped the man with a round from his M1.

The guns rattled and blazed for a half hour, while the mortarmen set up their tubes and lobbed shells on enemy positions higher in the hills. Lee skulked among the rocks, yelling in Chinese, trying to convince the last holdouts to surrender. Some emerged with their hands raised. Others, taking cover behind boulders in ones and twos, put up a stout fight. In the shadows, one Chinese soldier who knew a little English tried to masquerade as an American. He kept yelling, "No shoot, no shoot! I am Marine!" But it was only a trick to buy time. A moment later, he appeared from behind a rock, firing his weapon. The man immediately fell in a hail of Marine bullets.

It was past one o'clock when the firefight dissipated and the column was able to resume its forward movement through untrammeled snow. The men were spent, and yet they still had more than two miles to go. "We were like a chain gang of zombies," said one. "Exhaustion was telling on us," wrote Joe Owen. "Our falls became heavier, and it took longer for us to pull ourselves up from the snow." Even Lieutenant Lee had to concede that he had come to the end of his endurance. "I had to will myself forward," he said. "My thighs were like pillars of lead. How easy it would have been for an enemy soldier to have toppled me backward merely by tapping me on the chest."

The Ridgerunners stumbled forward through the early-morning hours, angling toward a massif that loomed in the distance. If Lee was reading his map correctly, this mountain was the final obstacle—just beyond it lay Fox Hill.

34

THIS PLACE OF SUFFERING

North of Fox Hill

In the gelid hours before dawn on December 2, the Ridgerunners tramped along the shoulders of Toktong-san, steadily advancing despite constant harassing fire from the Chinese. Colonel Davis hoped to reach the pass by first light, but he had not been able to establish radio contact with Fox Company—Davis wasn't even sure whether Captain Barber had been told that relief was on the way.

Then a sniper's bullet snapped through the morning gloom. It struck Colonel Davis in the head, knocking him flat on his back. Corpsmen bounded over to help, but Davis rose to his feet. He insisted he was fine. Upon examination, it was learned that the bullet had ripped through his parka hood and left a divot in his helmet. A scrap of shorn metal had grazed his forehead, but, miraculously, this was the full extent of his wounds. Davis brushed it off with his typical stoicism: Other than being shot in the head, he said, "all seemed relatively well."

Next it was Chew-Een Lee's turn. High on the flanks of Toktong-san, Lieutenant Lee encountered the same Chinese forces that had been attacking Fox Company since the night of the twenty-eighth. In the ensuing fight, a bullet found Lee. It struck the flesh of his already crippled right arm, near the shoulder. Lee crumpled to his knees. Though not seriously injured, he was in excruciating pain. He hissed invectives at the Chinese who had hit him, mocking them in their own language. A few Chinese soldiers emerged from their holes for a closer look, but Lee's men promptly mowed them down.

Davis was grateful that Lee hadn't been grievously wounded. The battalion had already suffered too many losses on the long night's trek. More than a dozen Marines had been injured seriously enough to require litters. At least three men had been killed in combat. Davis had ordered them to be buried in the snow. He would later rue this decision, but he felt he had no choice.

Then Davis was presented with a very different sort of casualty: One of his Marines had cracked. He was a strapping eighteen-year-old private from Texas, a kid who'd seen far too much. He was skittish, nearly paralyzed with anxiety. "Not going any farther" was all he said. At first people thought he was joking. The older men tried to buck him up. One Marine offered him chocolate. "We'll be back in Japan soon," another reassured him. But the kid was inconsolable. Nothing could break through his despondency. Refusing all overtures, he dug in his heels. "Not going any farther," he repeated. The battalion surgeon, Dr. Peter Arioli, examined him and could find nothing wrong. He babbled and shook, but he didn't appear to be wounded. He wasn't frostbitten and didn't seem to be shell-shocked. It was as though he had given up. Said one Marine in his squad: "The spirit had gone out of him."

Davis didn't know what to do. He couldn't leave the young man out here to die. Dr. Arioli declared him non compos mentis and had him physically restrained. They improvised a straitjacket and strapped him to a stretcher. Carriers took turns hauling the troubled kid across the mountain toward Fox Hill.

≡

Through the feeble light of dawn, the battalion fought its way up what Davis believed was the last spur of the last ridge: By his reckoning, as soon as they reached the top, they would see Fox Hill spread out before them. But Davis was wary of this final approach. He knew that it would be perilous to draw any closer to Fox without first making radio contact with Captain Barber. If Davis and his men were mistaken for the enemy, Fox Company's mortars would tear them to pieces. (After five days of besiegement, Barber would concede that his mortar and heavy machine gun teams were, as he put it, "getting pretty trigger-happy.")

Davis's radioman had trouble raising Fox on his hand-crank set, but after a few frantic minutes, he finally got a staticky signal. "Colonel, I have Fox Company," he shouted, nearly quivering with emotion.

Moments later, Barber's voice crackled over the waves. "This is Fox Six—over." The captain's businesslike tone brought tears to Davis's eyes. Barber wasn't surprised to be hearing from Davis; Litzenberg, in Yudam-ni, had already radioed Fox to say that a whole battalion would be tromping overland through the night. Still, Barber couldn't have guessed Davis's precise angle of approach, nor did he have, until now, any idea whether the Chinese had stymied his progress—or halted it altogether—through the long march. Davis could hear relief in Barber's voice.

"Fox Six," Davis said, "we are approaching the ridge preparatory to entering your sector. We will show ourselves on the skyline in five minutes. Will you alert your people to that effect?"

This was a roundabout way of saying *Hold fire.* Barber confirmed that he'd received the message loud and clear.

A few minutes later, when Davis had gained the top of the ridge, his radioman reached Barber again. "Fox Six, can you see us?"

"Yes, we can see you," Barber affirmed. Across the hill, the men of Fox Company crawled from their pits and hideouts to salute the figures assembling high on the ridge. Many of Barber's men produced scraps of parachute cloth, and the snowfields were aflutter with streamers of blue, yellow, and red.

Then Barber broke in: "Stay right where you are—I'm going to send up a patrol to guide you in."

Davis could only smile as he considered the bold incongruity of what Barber had suggested. Who was rescuing whom? Davis's Ridgerunners were an entire battalion; they didn't need a few bedraggled men from Fox to escort them the last few hundred yards. Davis was well equipped to fight his way in. But he appreciated the gung-ho sentiment behind Barber's gesture. It was as though the beleaguered captain were saying, *We're too proud to be saved.*

"Negative on that, Fox Six," Davis replied. "Keep your men in position."

≡

As Davis and his men marched onto Fox Hill, Corsairs appeared in the skies to cover one flank while Barber's mortars covered the other. In this way, a safe path was established for the Ridgerunners finally to funnel onto the field and unite with their embattled comrades. Davis's men would never forget the scene, simultaneously macabre and heroic, that unfolded before them. There was nothing beautiful about it—the hill was covered in blood and shit and trash and piss and guts and casings and chemicals. But it was as though they had entered a kind of Valhalla, a hallowed place in which every inch of ground spoke of some profound strife and sacrifice. Lieutenant Joe Owen described it eloquently. "We stood in wonder," he wrote. "Men bowed their heads in prayer. Some fell to their knees. Others breathed quiet oaths of disbelief." Tears welled in their eyes as they contemplated the savage ordeal that had transpired in "this place of suffering and courage."

What had happened here was remarkable to contemplate. A single company of surrounded men, outnumbered ten to one, had held on in arctic weather for five days and five nights. And not only held on: They had slaughtered their foe. The magnitude of the carnage filled the Ridgerunners with awe as they marched down through it. The field was littered with hundreds and hundreds of Chinese corpses. "I swear to God," said Owen, "you could have walked without touching the ground, using those bodies as a carpet." The faces on many of the Chinese dead, he said, "were frozen in spasms of pain."

These soldiers had died honorable deaths; their bravery was beyond question—nearly every one of them had fallen facing toward, not away from, the Marine positions. The far reaches of the slope had been cratered by shells. Chinese bodies had been stacked four and five high around the Marine foxholes.

Another Ridgerunner, radioman Joseph DeMonaco, thought Fox

Hill resembled "a Hollywood battle set." As they got closer to the Fox lines, DeMonaco started to see dead Marines. "One I'll never forget," he said, "was this Navy Corpsman, lying there with his scissors in one hand and a roll of bandages in the other. He must have been hit just as he was going to treat some wounded Marine."

As the Ridgerunners approached the perimeter, someone gave a raucous cheer, and they ran to join their comrades from Fox. According to one Marine account, "They sprinted through the snow with a speed that seemed to mock the awful night that had passed, and burst into Fox Company's lines with shouts and grins." The raccoon-eyed defenders emerged from their gruesome barricades of dead bodies to greet Davis's men. They wore raggedy slings and blood-drenched compresses, their heads wrapped in gauze. "Men hobbled about with makeshift leg splints," Owen wrote. "We exchanged profane greetings that did not conceal the love that we Marines felt for each other."

You look like shit! they yelled to one another—which, of course, was an insult of endearment. As one Marine put it, "The saviors would gaze upon the saved, and both would conclude that the other was pretty beat."

One of the first of the "saviors" to stumble into the perimeter was Lieutenant Chew-Een Lee, still in his unmissable costume of fluorescent panels. Lee had never been so exhausted in his life, and never so ecstatic. "It was exhilarating," he later said. "I was so proud of my men." He hesitated to say that he and the rest of Davis's men were "rescuing" Fox Company—although, practically speaking, that may have been true. "We never claimed that we saved Fox," he said. "Some of them may have resented us for this notion. But we relieved them from further attack—and we secured the pass."

Colonel Davis made his way to Barber's command post. Davis still wore his bullet-dented helmet, his forehead splotched with blood. The captain lay on a stretcher but was determined to greet Davis standing up. Wincing, Barber staggered to his feet, propping himself with his cane fashioned from a tree limb.

Ignoring each other's battle stench, the two officers smiled and warmly shook hands. At first they were too overwhelmed with emotion to speak. With the First Battalion's 450 men flooding onto the hill, there was no question that Fox would hold. What was more, they now stood an immeasurably better chance of breaking out of Toktong Pass and reaching Hagaru—and then the safety of the coast. Though Davis and Barber had much to be proud of, they were in no position to celebrate. Between the aid tents lay a pile of two dozen dead Marines, and more corpses were being brought in from the morning's fighting. Those bodies, stacked in the open air for all to see, chastened any impulse for rejoicing.

Barber went over the dour numbers with Davis. The men of Fox Company had sustained 118 casualties: twenty-six dead, eighty-nine wounded, and three missing. Of the company's seven officers, six had been hit—a few of them several times. Nearly everyone suffered from some degree of frostbite. With fewer than a hundred "effectives" left in its ranks, Fox could only with exaggeration call itself a company any longer.

≡

Davis's arrival may have drastically improved the strategic picture on Fox Hill, but it did not mean an end to the casualties. Although the Chinese had stopped their mass attacks out of deference to the American planes hectoring overhead, all through the morning, while Barber and Davis conferred, sniper fire kept angling through the camp. Every few minutes, another bullet came whanging through the air, splitting tent fabric, glancing off metal, spurting in the snow. Sometimes the Chinese bullets hit flesh.

Immediately upon arriving on Fox Hill, Dr. Peter Arioli, the surgeon, had found himself nearly overwhelmed by the task of taking care of casualties. Fox had some very brave and talented medical corpsmen, but to have a real doctor in the company's midst was a glorious amenity. One of the patients Dr. Arioli had to attend to was the eighteen-year-old Texan who said he'd had it and wasn't going to take another step. The straitjacketed kid was carried into one of the aid tents on a stretcher. His condition had deteriorated throughout the early morning. Arioli examined him but still could find no visible injuries. Then, a few hours later, to the incredulity of everyone, the young Marine passed away. The cause of death was never determined. He just stopped living.

"We were stunned," said one of his squad mates. "We carried him over to where the dead were, and put him at the end of the row. I think the poor son of a bitch was literally frightened to death."

Dr. Arioli was also supposed to be tending to Hector Cafferata, who had never received anything more than palliative treatment for his horrible wounds from five days earlier. The New Jersey private was hanging on, but he could barely breathe—his punctured lung was filled with blood. According to the clipboard, Cafferata was supposed to be next in line, but Arioli was detained on more urgent errands of triage. He had poked his head out of the tent to speak to someone, his gloved hands red with gore, when a sniper's round struck him. As described in one account, "A single bullet pierced the thin canvas and snapped Arioli's spinal column. The doctor was dead before he fell into the arms of the stunned staff officer."

"He didn't suffer," recalled a lieutenant who was standing nearby. "He was gone in seconds. One of the corpsmen examined him, and pronounced him dead." Dr. Arioli's body was added to the stack of corpses between the aid tents, next to the young Texan.

Another patient Dr. Arioli was supposed to have treated that morning was Chew-Een Lee. But because the lieutenant's arm wound was not deemed life-threatening, he was far down the list. Lee was resting in a warming tent when one of Davis's officers asked him to come interrogate a couple of Chinese POWs. Lee cut a sour look. He was not some egghead translator from intelligence. He'd spent the night leading a body of men through a mountain wasteland; he wasn't going to demean himself by serving as somebody's interpreter.

But Davis's officer prevailed upon him—these two Chinese POWs were kids, he said. They were, said one Marine account, "docile little fellows, sitting patiently in a grove of trees, hugging their knees in the wind." What they must have felt, as forlorn captives on a hillside that had become a public abattoir for more than a thousand of their frozen countrymen, no one could guess. But Lieutenant Lee went over and hunkered with them. They spoke with him freely, and Lee became engaged in their story.

"These people were former Nationalist soldiers who had fought under Chiang Kai-shek," Lee later said. "They did not parrot the usual boring Communist propaganda." What really motivated them, he learned—and, according to them, what motivated their most devoutly Communist comrades, too—was the sense that they were "defending the border, the frontier, from the aggression of foreign imperialists," Lee said. "You heard hardly a word about Communism, or Stalin, or Lenin, or even Mao. They considered our coming to the border a genuine threat to the motherland."

Lee asked them if they had any motivation beyond that. They did not hate Americans—as former Nationalists, they knew that the United States had long supported their cause. Why were they fighting Americans now? Why had their own leaders sent them here to die on this icy hillside, in the barrens of North Korea?

"*Mei yu fatzu*," one of them replied. It was an idiomatic expression used throughout China. It meant, basically, "We don't know. It can't be known. It's out of our control."

BOOK
FIVE

TO THE SEA

"The enemy you see before us is all that stands between us
and the sea we have marched so far and fought so hard to reach."
—XENOPHON, *Anabasis*

35

ATTACKING IN A DIFFERENT DIRECTION

Hagaru

The time had come for Oliver Smith to collect his regiments and start the process of chipping his way out of the icebox. All illusions of forward movement had been cast aside. They would be heading on, heading down, heading back, heading out. Quitting the field, if you wanted to call it that, but in a grand way, a planned way, engaging the enemy with every step. By radio, Smith had been following the movements across the battlefield, and today was the day when all the components of his master plan would begin to click into place.

First, Fox Company. He had heard about Davis's overland rescue mission, and he was immensely pleased to learn that it had been a success, that Barber had been relieved and Toktong Pass was again in American hands.

Next, his regiments at Yudam-ni, the Fifth and the Seventh. Both of them—some eight thousand Marines in all—would be breaking out that morning and marching toward him. They would be taking their wounded, their equipment, and some of their dead. Disengaging from their defensive posture would be a perilous thing, like letting go of a tiger's tail, but Litzenberg and Murray had it choreographed. They would advance as a "rolling perimeter," with companies of flankers thrown onto the ridges to clear the highlands while the main column of trucks, packed with casualties, eased along the valley floor. Weather permitting, squadrons of Corsairs would bomb and strafe the route ahead to keep the enemy scattered. The dash, all fourteen miles of it, was going to be bloody. No doubt the Chinese would hurl everything they had at the withdrawing Americans, but Smith thought Litzenberg and Murray had enough firepower to blast through. When they reached Toktong Pass, they would pick up the battered remnants of Fox Company, as well as Davis's Ridgerunners, folding them into their advance for the final miles to Hagaru. All in all, it seemed to Smith like a sound plan.

Finally, there was the Hagaru airstrip. After twelve days of non-stop effort, the engineers were reporting that the runway was 2,900 feet long—impressive, but still a thousand feet less than the length recommended for the big transport planes General Smith hoped to bring in. But Smith declared that it was time to stage the first test flight. The Navy surgeon at Hagaru had told him that the casualties had mounted to more than eight hundred, and he expected hundreds more when Litzenberg and Murray came in. Smith had to start evacuating the wounded now. He couldn't wait for the airstrip to reach regulation length. So he radioed a request for a plane.

At two thirty that afternoon, an Air Force C-47, piloted by a brave airman, dropped over the ridgeline and barreled in for a landing. The big twin-propeller plane hopped a few times on the bumpy airstrip, then jerked to a stop. Crewmen loaded the plane with a few dozen of the most seriously wounded. Now came the real test: On this short strip, in these cold temperatures, at this high altitude, would the C-47 make it off the runway and climb fast enough to clear the tight hem of mountains?

Colonel Partridge's engineers stood along the edge of the runway, beside their bulldozers, and anxiously watched. The plane jittered down the strip. It strained a bit, but the C-47 vaulted over the ridge with a little room to spare. Snarling into the sky, it banked for Hamhung as the engineers cheered.

Five more transports came in that afternoon, each one egesting more supplies and carrying off more of the reservoir's critically injured. The next morning, at first light, the process began again. Smith said the airstrip was starting to look "like La Guardia." Every ten or fifteen minutes, another plane would rumble in from the coast. The operation was taking on something of the quality of the Berlin Airlift.

The crates of ammo, the barrels of fuel, the pallets of C rations were starting to pile up at the edge of the runway. Smith's foresight was paying huge dividends. Hagaru was coming to life again.

And the aid tents were emptying. "What casualties?" Almond had asked a few weeks earlier. By the time the rescue flights drew to a close several days later, more than four thousand injured men would be spirited to safety. Smith couldn't be more pleased. The big birds kept coming, and the tent hospitals at Hamhung kept receiving the injured. With each flight, the forces at Hagaru grew leaner, the perimeter tightened, the enclave sharpened.

The surgeons in Hagaru's hospital were relieved now to have the option of diverting their worst cases to the evacuation planes. The doctors had been working around the clock in unbelievably trying conditions. Hagaru's hospital, set up in an old school building, had a gaping bomb hole in its roof that let in blasts of icy wind and snow. It was so cold in there that the doctors had trouble getting their plasma and IV drips to flow—the fluids had turned to slush. At times, the line between active surgery and active combat appeared dangerously thin. Bullets often came ricocheting through the operating room, pinging off medical equipment.

On one occasion, a half-crazed Marine rushed in, straight from the battlefield, holding a live grenade. It turned out the Marine had accidentally pulled the grenade's pin, and he was now nervously clutching its "spoon"—the only thing preventing it from exploding. The Marine had come into the busy operating room thinking that a bit of well-placed medical tape would finally enable him to loosen his grip on the deadly thing. A coolheaded orderly steered the Marine outside and convinced him to toss the grenade into a barren field, where it safely detonated. "You constantly felt the presence of the battle just outside our doors," said Dr. James Stewart, a recent Tulane Medical School graduate who was serving as one of the only surgeons on duty. "We were stretched to the maximum, but the airstrip considerably eased the pressure."

=

The completion of the airstrip also brought in the outside world. On some of the transport planes were stowaway journalists searching for a story. Perhaps the most famous among them was Marguerite Higgins, of the *New York Herald Tribune*. She arrived from Tokyo shortly after the airstrip opened and, as one of the first independent eyewitnesses to the events at Chosin, succeeded in getting some of the first dispatches out to the world. Higgins was as beautiful as she was resourceful, and the men, who had not seen an American woman in many months, remarked on how wonderfully incongruous it was to see her in this godforsaken place—though most were too tired and shell-shocked to summon prurient thoughts. Higgins described the Marines at Hagaru as men who'd been granted clemency. "They had the dazed air of men who have accepted death and then found themselves alive after all," she wrote. "They talked in unfinished phrases. They would start something and then stop, as if meaning was beyond any words at their command." Higgins wondered if they could possibly muster the strength to make the final punch to the sea. "They were drunk with fatigue," she said, "and yet they were unable to shrug off the tension that had kept them going without sleep and often without food."

Time-Life's David Douglas Duncan, armed with two Leica cameras, dropped into Hagaru on one of the flights. Some of the photographers had found that their cameras wouldn't work in the extreme cold, but luckily Duncan's equipment had been thoroughly winterized back in New York—all the oily lubricants removed so that nothing could freeze up. He immediately went to work, snapping what would become some of the most iconic images of the Chosin Reservoir campaign. But he was able to shoot only a handful of rolls; often the film would break when he tried to load his cameras. "I had to be extremely careful in that cold," Duncan said. "The film was so brittle, it would snap like a pretzel."

From the arriving journalists, General Smith began to learn how the battle was being portrayed in publications in America and around the world. Some newspapers had reported that Smith's division had already been destroyed, or at least that it appeared unlikely that he could ever spring his men from the "Red Trap." Headlines screamed that the Marines were a "lost legion," "surrounded," "doomed." Peking Radio announced that the "annihilation of the United States 1st Marine Division is only a matter of time." Hawkish politicians back home were urging Truman to drop atomic bombs on Chinese cities. The chief of the CIA, Walter Bedell Smith, worried that the Marines at Chosin could be wiped out unless the president began immediate high-level negotiations with Mao. Radio commentator Walter Winchell had called on the nation to hold a prayer vigil for the Marines at Chosin. "If you have a father, brother, or son in the 1st Marine Division," Winchell said, "pray for him now."

Some of the newly arrived reporters managed to persuade the usually reticent General Smith to say a few words. When Smith described how they would be driving to the sea, a British journalist interrupted him: "So you'll be retreating, is that it?" As the story goes, the general sternly contradicted him, saying, "Retreat, *hell!* We're just attacking in a different direction." Smith later denied that he'd said it quite this way, but the quote caught on as reported and would live on as Smith's most famous utterance. The more nuanced point that Smith sought to make was that he anticipated that their march to the coast would be a battle the entire way. When you are surrounded by overwhelming numbers of enemy soldiers who are trying to kill you, movement in any direction becomes, by definition, an attack.

In any case, Smith said, they were going to save themselves. They were going to come out with their units intact, with most of their equipment, and with all of their casualties accounted for. And when it was over, the First Marine Division would still exist as a formidable force, ready to fight another day.

≡

All this time, as the airstrip reached completion and the planes began to arrive, Hagaru remained engrossed in a full-scale battle. The Chinese had never given up on seizing the town. Night after night, their assaults on the perimeter were unrelenting. They were keen on taking the airstrip, which they viewed as a tremendous insult and a threat. During the day, Chinese snipers took potshots at nearly every incoming and outgoing flight. Though they never succeeded in downing any aircraft at Hagaru, they riddled the underside of many a plane. Patients aboard the rescue transports took the occasional enemy bullet while soaring over the Chosin battlefields. One evacuee reportedly was killed in this way, and a number of crewmen were shot. One of them was Robert Himmerich y Valencia, a seventeen-year-old Marine radio operator, whose feet were seriously injured when Chinese antiaircraft rounds burst through the bottom of the DC-3 transport on which he served. "Luckily, no control cables or hydraulics were hit," he later recalled, "but the plane was nicely peppered, which ventilated things a bit." Himmerich y Valencia lost the use of seven of his toes, but he survived.

On the ground, meanwhile, the battle for Hagaru did not let up. East Hill continued to be the focal point. Its broad shoulders saw some of the most savage fighting of the whole Chosin campaign. Sitter's George Company and Drysdale's Royal Marines had valiantly plugged holes and repelled the worst attacks. It was frenzied, all-or-nothing combat, much of it hand to hand. The George Company Marines, who cut strips of red parachute silk to wear as identifying scarves, became known as "Bloody George." Control of the hill changed by the hour, and the slopes became littered with so many Chinese corpses that the Marines employed them as windbreaks.

"Death was all around us," said Bob Harbula, a George Company Marine from the Pennsylvania steel country. "It was like a shooting gallery up there. Killing all those people—I felt like a mass murderer. Often we didn't have time to reload, so we wielded our rifles as clubs. Other times, we used our entrenching tools, even our helmets, as weapons." Standing in the town with a pair of binoculars, one could look up and see stacks of bodies on East Hill; some of them had become such prominent features that mortar teams began to use them as orientation points. But the Chinese, some of their officers on horseback, kept leading new assaults and reclaiming various parts of the hill.

When the mortar teams ran low on ammo, a supply drop was called in, using the established code name for 60-millimeter mortar shells: "Tootsie Rolls." A plane came and dropped the requested supplies by parachute, but when the boxes were cut open, the supply crews found no shells inside. Whoever had packed the order, apparently not familiar with the code name, had stuffed the boxes with actual Tootsie Rolls, enough candy for many thousands of men. The mortar teams were infuriated about the mix-up, but everyone else was thrilled. From then on, Tootsie Rolls became the signature treat of the Chosin campaign, and a form of currency. Many Marines would insist that Tootsie Rolls had sustained them through their darkest hours—and may have saved their lives. (The Chinese were crazy about them, too: When looting American positions, Tootsie Rolls, along with Marlboro cigarettes, were the first things they hunted for.) The Marines found that the confection not only provided quick fuel; it had a practical use, too. The rolls were the perfect size and consistency for plugging bullet-riddled gas tanks, fuel hoses, and radiators. The men would warm the candies in their mouths until they softened, then "precision-mold" them to whatever shape was required. Tootsie Rolls thus became a kind of all-purpose spackle that kept the Marines and their machines going through the battle.

Just when it seemed as if Smith might lose the battle for East Hill, the incoming planes paid off again: They began to bring in hundreds and hundreds of Marines from the coast. Some had been recovering from wounds sustained earlier in the campaign. Others were non-combat Marines—drivers, clerks, communications people, logistics people. But they knew their brother Marines were surrounded at Chosin and were eager to help. Their arrival was like an infusion of fresh blood. As they disembarked from the planes, Smith wasted no time in putting them on East Hill to stiffen the lines.

≡

The next day, December 3, the tide of the battle unmistakably turned. It was then that the Marines from Yudam-ni began to trickle into town. They came rolling in like some army of Joads, shabby but spirited, beat but not beaten, their ice-barnacled vehicles shot to hell, windows shattered, windshields cracked. The wounded were crammed into those same trucks—hundreds and hundreds of them—

while the dead were strapped to bumpers, tied to fenders and hoods, stacked like kindling on the roofs of cabs. And in truth, most of the living looked nearly dead themselves. "Our parkas were all stained with blood, food, gun oil, and dirt," wrote Lieutenant Joe Owen. "Our filthy faces were matted with bristly beards that bore icicles of mucus and spittle." Many discovered that they'd been sleepwalking the final few miles; once they arrived in Hagaru, said Owen, they collapsed in heaps and "slept dead to the world."

All day and night, the vehicles kept streaming in—troop trucks, tractors, ambulances, bulldozers, rolling artillery. The quick and the dead were mixed together in one long procession. More than eight thousand men in all. Hector Cafferata was in that procession, stuffed in a truck, holding his Mauser, one of his lungs clogged with blood. Bill Barber was in it, too, and Ken Benson, and so many of the casualties from Fox Hill. The worst cases were taken straight to the airstrip, where the triage doctors affixed them with evacuation tags. The less seriously wounded would have to wait their turn. The still less seriously wounded were handed weapons and sent out to fortify the perimeter.

General Smith greeted his regiments with a fatherly mixture of pride and relief. With their arrival, he knew that nothing could stop his division now. Alpha Bowser was sitting with Smith in a hospital tent when this happy realization started to sink in. "Our troubles are over," Bowser said to Smith. "We've got it made. The Chinese don't stand a chance."

"Bowser," Smith replied, "those Chinese never *did* stand a chance."

When Lieutenant Colonel Ray Davis, Lieutenant Chew-Een Lee, and the Ridgerunners marched into Hagaru, they straightened their shoulders and fell into parade formation, their boots keeping cadence on the snow. Then they began to hum a familiar tune. Soon they were singing "The Marines' Hymn" at the top of their lungs:

Here's health to you and to our Corps
Which we are proud to serve;
In many a strife we've fought for life
And never lost our nerve;

If the Army and the Navy
Ever look on Heaven's scenes,
They will find the streets are guarded
By United States Marines.

A Navy surgeon, standing on the road and watching the spectacle, could only shake his head in admiration. A reporter heard him say, "Look at those bastards, those magnificent bastards."

Litzenberg's and Murray's men were fed hot pancakes and given a good rest, but Smith soon began to organize them for the coming breakout. Vacating Hagaru was going to be a complicated endeavor: Over the next few days, they would have to shrink their perimeter, close down the airstrip, strike the hospital tents, and destroy everything of value they couldn't take out with them—using thermite grenades to melt and disable equipment made of metal. They would have to assemble the vehicles and arrange the forward units in a great flying wedge. All of this while continuing to fight a nightly battle.

But one more urgent problem was brought to Smith's attention: The Army units east of the reservoir, originally under the command of Colonel Allan MacLean, had met a disastrous fate. Several days earlier, Smith had refused to help them in their hour of need—he simply couldn't spare the men. But now that Hagaru had been reinforced, he had to do what he could to save the last Army remnants stranded on the ice. Some survivors from the battles on the east side of the lake had begun to straggle into Hagaru. The stories they had to tell were beyond horrifying.

36

IN THE DAY OF TROUBLE

East of the Reservoir

The wounded men jounced on the back of the transport truck as it rolled south in the long, fitful convoy. Twenty-five casualties of war, lying on pallets of parachute silk, were crowded on the open bed of this rolling medical ward. Some had been maimed by shrapnel or riddled with bullets. Some had sucking chest wounds. Some were dumbstruck from battlefield concussions, their eardrums perforated. Whenever the truck stopped, the soldiers winced. Each jolt seemed to pull at their wounds.

It was early on the morning of December 1, and the sun's rays slanted through the naked trees that fringed the ridgelines, giving light, if little promise of warmth. Ed Reeves sat somewhere in the middle of the truck bed, in this cargo of Army invalids. He was zipped to his neck in his sleeping bag, trying in vain to keep warm. A private first class, attached to Company K, in the Thirty-first Infantry, Reeves was a gangly kid, nineteen years old, with a clownish smile and furry eyebrows that quivered like caterpillars on his forehead. He had grown up on a farm in Illinois and had worked in a succession of Chicago factories before enlisting in the Army in 1949, looking for discipline and adventure. In Korea, he'd found both.

The truck engine spluttered and coughed, then died in a frozen creek bed. Reeves and the two dozen other men scoured the hills for Chinese while mechanics coaxed the engine back to life. "Think we'll ever get out of this mess?" someone asked. Reeves said he thought they would. Whatever his faults, a lack of optimism wasn't among them.

The truck, trying to catch up with the rest of the convoy, came around a turn and ran straight into the enemy. The Chinese swept down from the hills and fired into the vehicles. Bullets winged through the air, sometimes drilling into the sideboards of the truck, sometimes clinking off metal, sometimes striking flesh. A man to the left of Reeves was hit in the arm. Another, to his right, was hit in the face. Reeves cringed lower in his bag, but there was nowhere to hide.

The truck began to slip off the road, trailing sluggishly toward a ravine. It took a while, with all the shouting and shooting and screaming, to figure out what the trouble was: The driver had been shot and was slumped, dead, over the steering wheel. The windshield was spattered with his blood.

A soldier walking on the road nearby took the initiative. He pulled the dead driver from the cab and assumed the wheel, guiding the truck back onto the road. But not two hundred yards farther on, he, too, was shot, and a third driver had to be found. It was obvious whom the Chinese snipers were targeting. Anyone who climbed into the cab of a truck knew it was a death sentence. Yet, throughout the train of vehicles, men kept volunteering to drive.

Luckily, the day was sunny and clear, which meant that Corsairs from the First Marine Air Wing could be unleashed on the enemy. When a firefight broke out somewhere ahead of Reeves's truck, a Corsair was called to the scene. A plane swooped down from the treetops, scattering the Chinese. The pilot dropped his canister, but the releasing mechanism got hung up for a second too long. The drum of jellied gasoline tumbled through the air. It exploded, its contents splattering in the vicinity of one truck and a group of American soldiers who happened to be walking alongside.

A nightmare unfolded in front of everyone. The napalm clung to the victims' skin and clothing—they couldn't get it off. The burning petroleum gel kept boring into their flesh. They flung themselves on the ground, writhing in terror and agony, some begging their comrades to shoot them, some marching straight for the hills, hoping the Chinese would put them out of their misery. Soldiers rushed over and rolled the victims in the snow but found they were helpless to save their dying countrymen. "It was so hot out there, we couldn't breathe," one witness, Private Robert Ayala, said. "Them guys, their skin just fell right off. They was all charcoal, like a cigarette."

"The smell was awful," said another witness to the incident, Harrison Ager, an Army truck driver. "They were miserable. The quicker they died, the better off they were." Nineteen men, having spent the week nearly freezing to death, were sizzled alive. The charred bodies continued to smolder long after they had stopped twitching.

The convoy resumed its forward progress, but a grisly disaster like this, so early in their slow trundle to the south, hung heavily on the men and shook their morale. Would they ever make it to Hagaru? Even Ed Reeves's optimism was shaken. He lay in his truck, feeling for the good men he saw dying around him, astounded by their bravery. He yearned to join the fight, but that was impossible. He couldn't hold a weapon. He couldn't walk. He couldn't even stand.

$$=$$

During the first night of the fighting, back on the night of November 27, Reeves had seen plenty of action. He had fought well, he thought, and by the morning of the twenty-eighth he felt a certain cockiness, a sense that fate was smiling on him. He had burns on his jacket from grenade shards, he had scratches from the underbrush, he had bullet holes in his jacket. But he was unscathed. A little later that morning, though, as he was standing beside a dilapidated farmhouse that functioned as his company's command post, a round from an antitank gun landed beside Reeves. The blast hurled him twenty feet through the air. His legs and arms were sliced with hot shrapnel. The explosion knocked the breath out of him and rendered him unconscious. When he came to, he realized that his legs were wrecked and bleeding profusely. He couldn't get up.

Some men fighting nearby hauled him into a farmhouse and set him on the dirt floor beside a dozen other injured comrades. There he lay for three days, unable to move. Through a crack in the doorway, he watched the fighting outside and could see the Chinese growing closer. He had probably been concussed in the blast, and every explosion outside made his head throb. His legs hurt terribly as well, but he could not find a medic to administer morphine. He plugged the worst of his wounds with wads of torn cloth—and suffered in silence.

For those three days, all he had to eat or drink was a little tea steeped in melted snow and a breakfast package of bran cereal. He savored the dry flakes, one by one, and let them slowly dissolve on the back of his tongue like a delicacy. His only other diversion was a pocket Bible he sometimes thumbed through. A devout Christian, Reeves would read aloud from the Psalms during the heat of battle. "And call upon me in the day of trouble: I will deliver thee." There are no atheists in foxholes, it is said; whatever their background, none of the other casualties cooped up in the farmhouse seemed to mind Reeves's recitations.

Outside, along the east side of the reservoir, the Army units were unraveling. Their original commander, Colonel Allan MacLean, was gone. A few days earlier, marching across an icy inlet toward a company of men who he thought were American reinforcements but turned out to be enemy soldiers, MacLean had been riddled with bullets and taken prisoner. (He would soon die of his wounds while in captivity.) MacLean's successor, Colonel Don Carlos Faith, was left to make sense of the situation. Faith was the man General Almond had awarded a Silver Star on the morning of the twenty-eighth, the man Almond had told to buck up and not let a bunch of "laundrymen" stop him.

Finally, on the evening of November 30, the word came that Colonel Faith would be attempting a breakout to the south. No succor was coming from the Marines at Hagaru, they heard, so Faith had determined that the only hope for the nearly three thousand soldiers trapped here on the east side of the Chosin was to smash their way out. If Smith couldn't break through to Faith, Faith would break through to Smith. It was the Army's last chance to save itself from certain destruction. Shortly after dawn on December 1, Ed Reeves and the other critically wounded who could not walk—more than six hundred men—were loaded onto trucks while demolition crews burned equipment and dismantled weapons to prevent anything of strategic value from falling into Chinese hands.

Later that morning, at Colonel Faith's order, the exodus had begun.

=

By midafternoon the breakout had descended into havoc. The vanguard, inching along in the face of roadblocks, blown bridges, and booby traps, had covered only a few hard-won miles. Machine gunners had expended their ammunition. Grenades had become scarce. Many of the trucks ran out of fuel. As the cold sapped the radio batteries, various sections of the convoy lost communication with one another and with Colonel Faith. It was a long, deadly traffic jam, as far as the eye could see. The unwounded were wounded, and the wounded were *re*-wounded. Men cowered under trucks, some unable to fight, some refusing to fight, others already thinking the fight was as good as over. Company integrity began to crumble: Men from different units became confusingly intermingled on the road. No one seemed to know who to take orders from, or who to give orders to. Throughout the wrecked cavalcade, military order was collapsing. The column had become, as one historian put it, "a struggling organism without its central nervous system."

The enemy, sensing vulnerability, grew bolder. They rivered down from the ridges. When the sun began to set behind the hills, dousing the road in a dreary blue light, the Chinese grew bolder still, for they knew that with darkness the Corsairs would have to return to their bases. The Red soldiers lunged closer, sallying right into the column of trucks, sometimes shooting at point-blank range or bayoneting men and dragging them off the tailgates.

The Americans were powerless to stop them: The Chinese were cutting the convoy to pieces.

Sam Folsom, a Marine pilot, happened to be flying overhead in a Corsair and witnessed it all. "It was heartrending, like a bad dream," Folsom said. "I watched the Chinese coming down from the ridgelines. I could see the Army guys, all strung out, cut off, overwhelmed." Folsom, a highly experienced pilot from World War II, became so desperate to help the soldiers that he did something uncharacteristically impulsive and reckless: He swooped down over the trees, dropping lower than he had ever tried to fly a plane. He picked out a cluster of Chinese soldiers and, aiming directly for their heads, tried to chop them up with his propeller. "I knew it was crazy, but that's how close I got, I was so full of rage." Folsom came within just a few feet of succeeding in his maniacal undertaking, but then, coming to his senses, he pulled higher into the sky. He realized there was nothing he could do. His Corsair wasn't armed—that day, he was serving only as an observer. "I could see what was going to happen down there," he said, "and I was helpless to stop it. It was feet against wheels down on that road—and the feet were going to win."

Ahead of Ed Reeves's truck, Colonel Faith tried to rally the troops, marching up the line of vehicles with a Colt .45 automatic pistol in his hand. When a few men started to head for the safety of the woods, he threatened to shoot all deserters. Then he discovered two South Korean soldiers who'd lashed themselves to the undercarriage of a truck. Evidently, they had hoped to ride to safety, said one account, "like Odysseus escaping from Polyphemus." Faith ordered the two young Koreans to rejoin the fight. When they refused, he shot them both.

The Chinese mounted another attack, and Faith was mortally wounded—a piece of shrapnel from a fragmentation grenade had cut into his chest, just above his heart. Aides propped him up in the cab of a truck, like a mannequin, perhaps to make it appear as though he were still in command. The dead Faith looked, said one historian, "rather like El Cid riding out to his last battle." But when this deception was discovered, the men stopped fighting. Shorn of leadership, with no one else rising up to take charge and with ammunition virtually spent, the units began to dissolve.

By this point, farther back in the column, Ed Reeves's truck was stitched with bullet holes. Its gas tank had been pierced multiple times, draining the last dregs of fuel. The truck wouldn't budge. The men sat in the late-afternoon gloaming, waiting for the worst to happen.

A small reconnaissance plane came over the ridge. It wagged its wings, then dropped a series of containers along the road. Someone near Reeves's truck retrieved one of them. Inside was a leaflet that read: ENEMY ON ALL SIDES. DON'T STAY ON TRUCKS. CROSS ICE TO OUR TROOPS SOUTH END OF RESERVOIR. A sergeant, marching through the column, came along to reinforce what the leaflet said. He stopped at Reeves's truck, climbed into the bed with the wounded, and told it straight: "Column ain't moving again. Everybody who can, better move out."

The men were confused. Move out *where?*

The sergeant pointed toward the reservoir, a ghostly sheen off to the north, a mile or two away. "Get out on the ice. And head left."

And if we can't move at all?

The sergeant was blunt. "Prepare to become prisoners of war," he said.

And so began the final unraveling. In ones and twos, then in clusters, then in throngs, the able-bodied men quit the road and started running, with the not too badly wounded plodding behind. They abandoned their trucks, they abandoned their dead; many even abandoned their weapons. What good was a rifle without ammunition? It would only slow them down. They ran in serpentine fashion as they made for the reservoir, dodging Chinese bullets. Officers tried in vain to instill some semblance of discipline. What had once been an army had devolved into a rabble.

$$\equiv$$

Ed Reeves unzipped his bag and tried to stand, but the pain in his legs was so severe that he started to faint. It was futile: He couldn't move an inch, let alone walk to the reservoir. The truck had been his only hope of deliverance. Now it had become his prison.

Reeves looked down the road and caught a play of shadows, an approach of human forms. Bracing for the worst, he zipped his bag and watched out of the corner of his eye. The figures walked up to the tailgate in the crepuscular light. Reeves sighed in relief: They weren't Chinese. They were North Korean civilians, a group of women and children. Would they harm us? he wondered. Rob us? The civilians stood for a few moments, looking sorrowful, studying the American wounded with solemn expressions. One of them stepped forward and uttered a few soft words that sounded to Reeves like a benediction. Then they bowed and turned away.

When night fell, Reeves fished out his pocket Bible and read by

(ABOVE) Advancing Marines halt as aviators provide close air support in the distant foothills. *Marine Corps History Division*

(RIGHT) A Marine rocket team moves along a bitter cold ridgeline near Chosin Reservoir. *Marine Corps History Division*

(BELOW) The morning after the first night's battle, the corpses of Chinese soldiers are strewn along the hills above Yudam-ni. *Marine Corps History Division*

(BELOW) Lieutenant Chew-Een Lee, the Chinese American platoon leader who trailblazed the overland route to relieve Fox Company *Marine Corps History Division*

(ABOVE) Ensign Jesse Brown
Naval Historical Center

(BELOW) Lieutenant Thomas Hudner
Naval Historical Center

(ABOVE) Lieutenant Colonel Ray Davis, leader of the Fox Hill rescue mission *Marine Corps History Division*

A squadron of Navy Corsairs, rimed with ice on a carrier in the Sea of Japan, await their next mission. *Naval Historical Center*

At Hagaru's newly completed airstrip, a transport plane prepares to evacuate the Chosin wounded. *Marine Corps History Division*

Marines at Koto-ri collect the frozen corpses of their comrades. *Marine Corps History Division*

An Army ambulance, riddled with Chinese bullets during its retreat from the mountains *Marine Corps History Division*

"Attacking in a different direction": The Marines begin their withdrawal to the sea. *Marine Corps History Division*

Exhausted Marines during a pause in the exodus *Marine Corps History Division*

A Marine, marching in solitude, is enveloped in snow. *Marine Corps History Division*

(ABOVE) Marine engineers inspect the Funchilin Pass bridge after it was blown by the Chinese. *Marine Corps History Division*

(RIGHT) Engineer Lieutenant Colonel John Partridge, chief architect of the bridge repair *Marine Corps History Division*

(BELOW) Thousands of North Korean civilian refugees await departure from the port of Hungnam. *Marine Corps History Division*

(ABOVE) Evacuating Marines board an amphibious vessel at Hungnam. *Marine Corps History Division*

(LEFT) Jesse Brown's widow, Daisy, looks on as President Truman awards Tom Hudner the Congressional Medal of Honor in 1951 during a White House ceremony. *National Archives*

(BELOW) A bugler plays taps before freshly dug American graves at the Hungnam cemetery. *Marine Corps History Division*

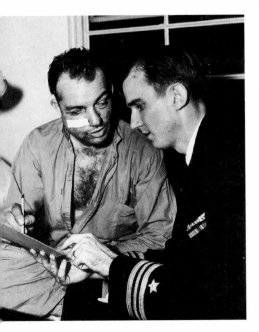

(ABOVE) Lieutenant John Yancey, recovering from facial injuries in a Japanese hospital, recounts his battle experience to a Navy representative. *MacArthur Museum of Arkansas Military History*

(BELOW) Dr. Lee Bae-suk, with his wife, Mi-yong, at their home in Cincinnati in 2017 *Hampton Sides*

(BELOW) Private Ed Reeves, having recovered from his battlefield and frostbite wounds, photographed with his wife, Beverly, after their wedding in 1952 *Army Heritage and Education Center*

General Oliver Prince Smith, photographed after the Korean War *Marine Corps History Division*

moonlight from the Book of Psalms until he fell off into a fitful sleep. Shortly after dawn, a figure approached the truck. It was a solitary man, his boots crunching on the snow. He walked around to the back of the truck and stood at attention. The man appeared to be a Chinese officer. He wore a gray greatcoat and a fur-lined cap. "Good morning, men," he said in nearly perfect English. He seemed educated and poised, and he spoke softly, with a British accent.

One of the wounded sitting on the tailgate implored the officer. "Sir," he said. "You can see we're dyin' here. If we don't get shelter and medical help soon, we'll all be dead. Can we move to a place where we won't freeze?"

The officer stepped forward and raised his mittened hands in a gesture of impotence. "I am sorry," he said. "I can give you none of the things you ask for. If I were to be heard speaking to you like this . . ." He stood, looking sympathetically at the wounded men for a moment. "I have only stopped to say, 'Bless you all, may God be with you.'"

Then he spun around and marched up the road.

=

Reeves, having dozed through the middle of the day, woke with a start. He thought he heard gunshots somewhere ahead. "What's that?" he demanded. "What's going on up there?"

People in the forward part of the truck craned their necks to see. "Gooks are burning the trucks," someone yelled. The man's tone was strangely neutral, disembodied, as though the scene he watched and narrated were taking place on another plane, in another world—as though it didn't apply to them. "Looks like they're shooting anyone who tries to escape."

Finally it dawned on Reeves what was going on: The Red soldiers must have been given an order to clear the road, to dispose of the American wounded and dead. They were incinerating the trucks with the wounded still inside. Reeves could hear the screams of the Americans. He smelled gasoline and roasting flesh. The Chinese were being systematic about it, working their way back through the convoy. They would soak a truck with fuel, then toss in a few phosphorus grenades, then shoot anyone who found the strength to flee the flames. Then they would move on to another truck.

"Looks like we're next," someone said, adopting the same disembodied tone. "Here they come."

Reeves lay in a stupor, searching for some way to accept what was happening. When he joined the Army, he had imagined dying in any number of ways—heroically, in battle, performing some gallant deed. He never envisioned anything like this: being burned in a pyre with a few dozen of his helpless countrymen. As the Chinese approached the truck, he prayed: *God, give me the courage to die like a man.*

The Reds walked around to the rear of the truck. One of the Americans perched near the tailgate offered his cigarettes, attempting to make friends with his executioner. He was promptly shot.

Reeves could hear other Chinese soldiers rattling around the gas tank, apparently trying to siphon fuel with which to start the fire. But when they discerned that the bullet-riddled tank was empty, the officer in charge issued new orders: A soldier, armed with a rifle, climbed onto the bed of the truck. He came to the first man, leveled the barrel between his eyes, and fired. He moved on to the next. Then the next.

Reeves sat up, half out of his sleeping bag, and watched as though it were a movie. He didn't know any of these Americans especially well—he'd just shared the bed of a truck with them for a day and a half—but he was in awe of them. They didn't beg or cry; they didn't utter so much as a whimper. They looked their executioner in the eye and died with dignity.

When the barrel slid over to Reeves, he felt strangely at peace. His eyes followed the smooth metal until they found the eyes of the soldier. Reeves did not see malice in them. He was a kid, following horrible orders. Possibly, from his point of view, it was a mission of mercy—get it over with in an instant, rather than let them slowly freeze to death. The soldier was standing no more than three feet away when he squeezed the trigger. The muzzle blast knocked Reeves to the floor. Then the barrel swung over to the next man.

37

I'LL GET YOU A GODDAMN BRIDGE

Funchilin Pass, South of Koto-ri

As General Smith planned his breakout from Hagaru toward Koto-ri, a new problem presented itself further along his escape path to the sea. Aviators were reporting that the Chinese had blown the concrete bridge at Funchilin Pass, just below Koto-ri, making the road impassable. It was the same bridge Smith had been so apprehensive about when he'd first laid eyes on it, a few weeks earlier, on his way into the mountains— the precarious structure that spanned the penstocks carrying reservoir water toward the valley below. If the air reconnaissance reports were true, Smith knew, his exit plan was imperiled.

Smith had more than one thousand vehicles. They would form a convoy several miles long. These vehicles would transport the wounded and the dead and all the food and supplies his forces would need for their fighting exodus to the coast. As a matter of principle, Smith was determined not to abandon his equipment, nor to leave behind anything valuable for the Chinese. But now, apparently, his escape hatch had been sealed shut. The mood in Smith's headquarters was morose. His chief of operations, Colonel Alpha Bowser, viewed the situation this way: "My personal, private reaction when I heard the news was something akin to despair."

Smith again summoned his head engineer, Lieutenant Colonel John Partridge, to discuss the reconstruction project. Only a week earlier, Partridge had put the finishing touches on the Hagaru airstrip, a feat many people said couldn't be accomplished, and since then his teams had kept themselves busy repairing the road and clearing impediments along the MSR. Smith was starting to recognize that, to a peculiar extent, the battle of the reservoir was proving to be an engineering story. Now here was an even more nettlesome problem for Partridge's gifted mind to solve.

The two men talked for a while, then Partridge said he needed to see the destroyed bridge site firsthand. He crawled into a small observation plane and took off from the Hagaru airstrip, turning south over the snowy mountains. The spotter plane flew past Koto-ri and then bumped its way through the freezing air a few more miles until it came to Funchilin Pass. After he circled the site, the full import of what the Chinese had done began to register with him. This, said Partridge, "was a damned serious situation."

The Chinese sappers had turned the bridge into a mess of rubble and twisted rebar. It had a yawning hole, about twenty feet across. With the explosive force the Chinese had obviously employed, it was a wonder they hadn't ruptured the penstocks, causing a cataract that might have swept away the entire installation. Partridge could see that the mountainside to which the narrow road clung was too steep for him to construct a bypass around the ruined bridge—nor could he build an impromptu ledge on its downhill side.

Worse, Partridge discovered that the enemy had deposited another lovely gift a few hundred yards down the MSR, at a place where a trestle had once carried a set of railroad tracks up and over the road. Here, the Chinese had blown the trestle from its concrete pylons, neatly dropping the structure athwart the MSR. Partridge couldn't guess how many tons this huge mesh of metal might weigh, but he was worried it could be an even more significant obstacle than the blown bridge.

Partridge kept passing over the site in the little plane, examining it from every angle. He instructed the pilot to drop lower so he could get a better look. Only then did he see that thousands of Chinese were dug into the surrounding hillsides. Song's armies were preparing for a big fight here around the blown bridge. Every now and then, soldiers aimed their rifles in the air and took potshots at the plane. If Partridge flew any lower, he might be shot out of the sky.

Clearly the Chinese had recognized the vulnerability of this spot and had already taken efforts to exploit it. These were elements of the Chinese Sixtieth Division, and they had been given the order to do everything in their power to block the Marines' march south from Koto-ri. General Song, an avid reader of Sun Tzu, was doubtless familiar with one of the ancient philosopher's more widely quoted precepts: "When a cat is at the rat hole, 10,000 rats dare not come out; when a tiger guards a ford, 10,000 deer cannot cross." Funchilin Pass was the rat hole, and Song's Sixtieth Division was the cat. Here the Chinese would put up a determined fight. It was their last chance to prevent the Americans from escaping the mountains and slipping away to the coast.

Partridge shivered in the plane's freezing cockpit and tried to force his numb fingers to take notes. He knew there was neither time nor available materials to build a bridge of hewn timbers. Nor did he think the site would accommodate a Bailey bridge—a span that his engineers often used, constructed of prefabricated trusses. Partridge had to devise something different. It had to be foolproof and strong enough to withstand the weight of a Sherman tank, but also something that he could lay down in a hurry. After a few more passes over the site, he finally struck upon an idea.

≡

Upon his return to Hagaru, Partridge did a little research and sketched out some numbers, then met again with Smith in the general's headquarters, beneath the portrait of Stalin. Partridge had to inform Smith that the situation looked bleak at Funchilin Pass, that the topography was unfavorable, and that clearing the Chinese was going to be tough. Smith understood how thorny the problems would be. It was "the most difficult defile through which the division had to pass," he wrote. Smith had to give the Chinese credit. They "could not have picked a better spot to give us trouble."

The good news was, Partridge thought he could repair the bridge. He informed Smith that he would be constructing a modified version of what was known as a treadway span. This was a pontoon-style contraption—normally designed to float on the surface of a river or canal—that relied on portable sections of heavy steel girders bolted together at the site.

"A treadway?" Smith said, puzzled. He knew they didn't have any treadway pieces up here.

Partridge said he had already located some in Japan and could have them flown over straightaway.

"How will you get them up here?" Smith wondered.

"Air-drop them."

Smith was skeptical. Has this ever been tried before? he wondered.

"Never heard of it, General."

It was a bold solution—deus ex machina. The problem was that each treadway section weighed 2,900 pounds. No one seemed sure what would happen to a modest-size airplane, especially in thin mountain air, when that much heft was released at once. It might create such a dramatic shift in weight distribution that the aircraft would dangerously lurch and possibly spin out of control. Not many pilots were skilled enough or crazy enough to want to try it. Furthermore, no one was sure whether the treadway structures would survive the descent. Would they buckle and smash to pieces upon striking the ground? Partridge didn't know whether the military made parachutes large enough to sufficiently slow the fall of something so heavy.

Still, he believed the plan could work, and he was impatient to get started. His idea was to drop the treadway pieces inside a protected perimeter surrounding Koto-ri, where they could be retrieved and loaded onto large Brockway trucks—the Army happened to have two of these behemoth utility vehicles already in Koto-ri. The Brockways, which weighed six tons apiece and were designed for erecting bridges, would deliver the treadway sections to the site once Smith's forces had cleared the pass of most of its enemy defenders.

Employing multiple teams of laborers, Partridge would try to assemble the bridge as quickly as possible. He knew the task wouldn't be easy for his crews—constructing a bridge while simultaneously taking enemy fire—but they had worked under similar conditions while building the Hagaru airstrip. If Smith could keep the Chinese more or less at bay, Partridge thought his men could have it bolted together in a few hours. Once the long convoy had passed over the repaired bridge, a demolition team would then promptly destroy Partridge's handiwork, to deny the Chinese its use.

Smith listened intently to this. The plan had merit, he thought. But when he began to second-guess it, envisioning the things that might go wrong, Partridge, normally a tactful and even-tempered man, grew annoyed. "Dammit sir!" he snapped. "I got you across the Han River. I got you an airfield at Hagaru. And I'll get you a goddamn bridge at Koto-ri!"

Smith chuckled at the outburst. If he was startled to hear a lieutenant colonel cursing at a general, he didn't let on. He liked Partridge's improvisational spirit, liked the novelty and ambition of the idea. There was something classical about it, Smith thought. Partridge's solution may have been a long shot, but it was the only one Smith had. As one Marine account noted: "On such slender threads as bridges dropped by air do men and armies live or perish."

Smith ordered the engineer to get started immediately.

≡

Later that day, an Air Force C-119 Flying Boxcar growled off from Yonpo Airfield, near Hamhung, its belly loaded with a single treadway piece. The pilot had been ordered to rise over the rice fields, circle around, and then release the ponderous cargo within the confines of a drop zone marked with bright-orange panels. It was a hastily arranged rehearsal, under relatively controlled conditions, to learn what would happen. The crew successfully discharged the treadway piece, but the twenty-four-foot parachute attached to it—the largest one they could find in Hamhung—proved inadequate. The bridge section plummeted, crumpled upon impact, and was driven into the ground, sending up a cloud of dust.

The test had demonstrated that for the real drop at Koto-ri, Partridge would need much bigger parachutes. A special detail of Army rigging experts flew in from Japan and worked through the night, making calculations, coiling ropes, and packing chutes. At Yonpo Airfield, crews numbering more than a hundred men would labor until dawn, loading and readying the planes. Nothing quite like this had ever been tried before, and they were excited by the challenge. The rigging experts decided that each steel section should be dropped from eight hundred feet above the ground and would require two forty-eight-foot parachutes, one attached at each end. The riggers felt confident that this would be sufficient to break each piece's fall, but they were only guessing.

The next morning, a squadron of Flying Boxcars would be heading up to Koto-ri, each one bearing a precious piece of steel. Partridge knew he didn't have the luxury of staging any more tests. This time it would be the real thing.

38

BLOOD ON THE ICE

East of the Reservoir

Ed Reeves didn't understand how he could still be alive. His head pounded, and he felt hot blood oozing from his temple. He lay on the floor of the truck bed and tried not to move as he heard the rifle blasting his fellow Americans, one by one by one, into eternity. Another minute, and the shooting was over. The Chinese soldier hopped from the bed, and the officer in charge humphed his approval. The crew moved on to the next truck.

For hours, Reeves shivered in his bag. With Chinese soldiers all around, he dared not budge. Later in the day, though, when movements on the road quieted, he sat up long enough to steal a good look at the two dozen Americans around him. No one stirred. Their rigid limbs were twisted into awkward shapes, their faces frozen in odd expressions. He was sure he was the only one alive. He removed a glove and felt his face. Though it was tacky with blood, he could not find a bullet hole. The round had ripped a large flap of skin from his scalp and left a groove along his skull. But the gunman, in his haste to finish an unpleasant job, had fired a glancing shot.

Reeves didn't have a plan for what to do next, but he knew he had to get off the truck. The dead bodies gave him the creeps. When darkness came, he unzipped his bag and tried to stand, but it was just like the last time he had tried it: He felt so lightheaded, he thought he would faint. He realized that if he lost consciousness for any period of time outside the warmth of his sleeping bag, he would quickly freeze. So that was it: He had no options. He would have to stay the night here in this ghoulish place.

≡

The next day, Reeves was roused from his stupor by a crew of Chinese soldiers. They climbed onto the truck bed and started rifling through the clothes of the American corpses in search of anything valuable—watches, rings, cigarettes, lighters, boots. Once they were satisfied with their pickings from one corpse, they would toss the loot onto a cloth they had spread on the ground. They would hurl the corpse over the side, and other soldiers would drag it over to a fast-growing stack of bodies beside the road. Reeves understood what was happening: This was a graves detail. The Chinese were clearing the road of vehicles and bodies, foraging for useful relics, and trying to tidy up a hideous scene.

Reeves, peeking from his sleeping bag, tried to will his body not to tremble or make a sound. The crew worked down the line of corpses until they came to him. They unzipped his bag, unfastened his clothes, and started fumbling through his pockets. Reeves breathed shallowly and kept his body rigid, trying his best to assume the attributes of a frozen cadaver. But one of the Chinese soldiers, sensing human warmth, recoiled in horror. Reeves heard a gasp, then shouts. The next thing he knew, they had hurled him from the truck bed. He struck the frozen ground with a thud. Three soldiers picked him off the snow and leaned him against the tailgate.

Reeves opened his eyes wide—he was no longer playing possum. He sensed that they wanted him to stand up and bow in respect. When Reeves failed to do so, they started beating him, kicking him, slapping him in the face.

They dragged him over to the side of the road and hurled him onto the pile of corpses. Then Reeves saw them pick up their rifles. He was expecting a bullet. Instead the soldiers started pounding him in the head with their rifle butts. He tried to buffer the impact with his hands, but the blows kept coming. Several times he felt the bones in his hands and fingers cracking.

The soldiers kept smiting away until they thought he was dead. One of them grasped him by a hank of bloody hair and studied his face. Reeves stared vacantly down the road. He didn't blink. He didn't move. He didn't even breathe, fearing that his breath would freeze into fog as it rose through the air. The soldier let go of his head and grunted. He wiped the blood from his hands on Reeves's jacket. The crew seemed sure he was gone.

They crouched beside their loot. Through a blood-red veil, Reeves watched them haggle over some of their spoils. They folded up the fabric and spun it into a bundle. Then they vanished into the whirling snow.

=

After an hour or so, Reeves slid down from the pile of the dead. He tried to fasten his clothing back together where the Chinese had undone it, but his hands were so smashed up, he couldn't work the zippers and buttons and snaps. He sat in the snow and saw that the enemy was camped on the ridges above him. He was still in mortal danger—he had to get away from the road.

So he attempted to crawl, but his broken hands hurt so badly that he couldn't put any weight on them. He tried using his elbows instead of his hands. That seemed to work, and he was able to creep off the road and into a field. As he did, a patrol of a half-dozen Chinese soldiers came marching down the road. Reeves stopped crawling and looked back in dread. They couldn't possibly miss the trail of fresh blood that led from the corpse pile to where he was now. But if they did notice it, they didn't pay it any mind.

He resumed his pitiful crawl. He was in excruciating pain, and each bout of exertion would result in progress of only a few feet. He looked up at the ridge and saw hundreds, maybe thousands, of Chinese. He wondered, *Why don't they shoot me?* He would see them pop their heads out of their foxholes. It would have been so easy for them to snipe at him, but they held their fire. Reeves thought maybe they were taking bets on how far he could go before collapsing.

He came to a steep embankment leading to a set of railroad tracks that gleamed in the late-afternoon light. Painfully, he slid down and crawled across the tracks. He thought he heard the sound of marching feet reverberating through the cold metal of the rails. He climbed the other side of the ravine and hid in the scrub. A minute later, a column of Chinese tromped past him. They didn't see the splotches of new blood along the tracks; they didn't spot the freshly disturbed snow.

Hours passed, and finally Reeves reached the ice of the reservoir. He knew because he could feel its implacable hardness beneath him. Lying on his belly, he brushed away the snow and saw the blue glint in the moonlight. It was a beautiful sight, he thought, and for the first time he felt a glimmer of a chance that he might live.

He kept crawling farther out onto the reservoir, putting distance between himself and the Chinese. He found that his balky limbs responded to thought commands—individual parts would move only when he consciously willed them to. *Right foot, slide forward. Left knee, push off.* His elbows, no longer sturdy, would often slip out from under him and he'd bang his chin on the ice. At least it would jar him awake for a while. With redoubled determination, he would continue on through the night, scratching a jagged path across the long, smooth surface of the lake. To establish some kind of rhythm in his crawl, he recited the old marching chants he'd learned back in training:

> *Sound off!—one, two,*
> *Sound off!—three, four,*
> *The captain rides in a jeep,*
> *The sergeant rides in a truck,*
> *Gen'ral rides in a limousine,*
> *But you're just shit outta luck!*
> *One, two, three, four, one, two,*
> *THREE-FOUR!*

Counting cadence in this way worked for only so long. Reeves knew he needed to conserve his energy or else he was going to pass out from exhaustion—and freeze. So he angled his back to the wind and curled up in a fetal position, tucking his knees under him, mashing his hands into his armpits to keep them warm. He vowed not to doze off—he would simply rest for a short while. But a few minutes later, he heard boots squeaking over the snow. Reeves threw a glance over his shoulder and saw a Chinese soldier, holding a burp gun, sauntering up to him. The soldier appeared to be alone. Reeves couldn't figure out where he'd come from. The Chinese man prodded Reeves in the back with the barrel of his gun.

Reeves, who by this point was feeling nothing so much as a deep sense of disgust at the remorseless ways in which fate seemed to be toying with him, rose and cried out in a primal scream of spite and frustration. He figured that this time, surely, the end had come. But when he turned around, he was surprised to see the Red soldier running away. It was as though the man had seen a ghost. Reeves watched him kicking up snow as he sprinted toward the shore.

Reeves, wide awake now, continued his crawl across the ice.

=

At the first suggestion of morning, Reeves looked out and spotted another figure on the reservoir. He couldn't tell for sure, but he appeared to be an American GI. He had emerged from the brush of a peninsular shoreline and was stumbling along, hugging himself in a way that indicated he'd been injured in the gut. "Hey, you!" Reeves yelled. "C'mere and help me!" The man did come, and when he approached, the two invalids surveyed each other. Reeves thought the GI looked ashen, on death's door. God only knows what the man thought of Reeves—this sad creature, crawling along on his elbows like a praying mantis, his head a tangle of blood and ice. They stood there for a moment, looking at each other.

Then, out of nowhere, three Corsairs came zooming overhead, flying their first morning sorties. Reeves told the other soldier, "Write something in the snow—SOS or whatever." In a frantic fury, the man scraped his foot across the snow, etching out an *H*, then an *E*. He was on the long leg of the *L* when one of the pilots spotted the letters taking form below him. The plane came down low over the ice, so close that Reeves could see the aviator's goggles and the outlines of his helmet. The pilot gave them an "OK" signal as he flew by.

He must have radioed to dispatchers on the ground, for soon

Reeves and the other man could make out a jeep speeding across the reservoir. Reeves worked himself to a sitting position as the vehicle grew bigger and bigger on the ice. It pulled right up to him. A Marine colonel, whose jacket said BEALL, climbed out of the jeep. He squatted beside Reeves on the ice. "Soldier," he said softly. "Where you hurt? I don't want to hurt you more."

"The legs hurt a lot, sir."

Lieutenant Colonel Olin Beall, commander of the Marine First Motor Transport Battalion, gently gathered him and tucked him into the jeep. He removed his own fur parka and bundled Reeves up. As a second vehicle arrived to retrieve the other frozen GI, Beall climbed into his jeep and rapped on the side. "Ralph!" he barked. "Let's go!"

Private First Class Ralph Milton, a nineteen-year-old farm boy from Wyoming, was Beall's driver. For the past three days, at General Smith's instruction, they'd been pulling men off the reservoir. Beall's crews of rescuers, who had become known as the "Ice Marines," had saved more than three hundred men. Countless times, Beall and his drivers had risked their lives out here on the reservoir. Sometimes they'd heard an ominous creaking sound and realized that their jeeps were in danger of breaking through. Often they had drawn enemy fire as they pulled wounded GIs from the shore brush. The Chinese bullets would ricochet off the ice, sometimes kicking up spurts of water around Beall's crews as they sped back toward the Marine perimeter, using the frozen sheet as a superhighway.

Beall's team of rescuers, pulling alongside so many prostrate bodies on the ice, had to make dreadful, on-the-spot assessments about who was dead or alive. "We developed a technique," said Linus Chism, another of Beall's drivers. "We got to where we could tell by looking closely at the guy's mustache. If the hairs right under his nostrils was all icy and frozen, he was a goner. But if those hairs had even a little bit of moisture on them, just a little spot that wasn't frozen, that meant he was still breathing, and we'd pick him up."

Over the past three days, Beall and his drivers had seen many terrible cases. Men with compound fractures. Men walking in circles, muttering to themselves. Men hiding inside wooden dinghies that were locked into the shoreline ice. But he didn't think he'd ever seen anyone in such bad shape as Reeves. Private Milton drove as fast as he could, the jeep fishtailing over the ice. When they reached the Hagaru perimeter, the sentries waved them on through. "Head for the runway, Ralph!" Beall said. The jeep got out on the runway and made straight for a C-47 transport plane that was preparing to take off, its propellers turning. Milton flashed his headlights at the plane as he pulled around to its wing. Beall climbed out and pounded on the fuselage. The hatch opened, and the pilot emerged.

"I got another one for you here," Beall yelled over the roar of the engines.

"Colonel, we're already overloaded," the pilot said. "We can't take any more."

But Beall was adamant. "Captain, I said I got another one for you here." They loaded Reeves onto the plane and strapped him into a stretcher. The C-47 revved its engines and took off. As it pulled up over East Hill, Reeves heard an insistent metallic noise: It was Chinese small-arms fire, pecking the belly of the aircraft as it rose from the battlefield.

Reeves was one of the last soldiers to be pulled off the ice. When the final tallies came in, they were devastating: About a third of the nearly three thousand soldiers who were trapped on the east side of the lake had been killed or captured. Task Force Faith had been obliterated as a fighting force. In a way, Reeves was one of the lucky ones.

When the plane leveled off, a flight nurse leaned over Reeves. "Hurting much?"

"Yes, ma'am."

"When did you have your last shot for pain?"

"Haven't."

"When were you hit?"

"I don't know. Days ago."

She worked up a dose of morphine and administered it to Reeves. Feeling flushed and warm all over, he drifted into semiconsciousness. The plane buzzed toward the coast. It landed at Yonpo Airfield, near Hamhung, and Reeves was spirited inside a medical tent next to the runway. When they snipped his clothes away and beheld his wounds, the doctors didn't know where to start. His legs? His head? The doctors didn't say anything to Reeves, but they could see that he would have to lose his feet, and probably his hands as well. But Ed Reeves was defiant. "They give me another shot of morphine," he said. "By then I thought I could whip the whole Chinese army."

39

TAKING DEPARTURE

The Sea of Japan

On the afternoon of December 4, the aircraft carrier USS *Leyte* heaved in the seas off Hamhung-Hungnam. On its wooden flight deck, a pilot, Ensign Jesse Brown, revved the engine of his Navy fighter, Corsair 211, and prepared to take off. Crewmen removed the chocks, and the dark-blue plane pulled forward in a crowd of other Corsairs from Fighter Squadron 32. It was forty degrees on the carrier deck—brisk and gusting, but a world away from the mountains around the reservoir, where the squadron would be heading today. As the prop wash streamed over his plane's Plexiglas canopy, Brown cinched his helmet, adjusted his goggles, and began a final check of his instruments. He wore fire-retardant gloves, a khaki flight suit, a parachute, and a life jacket. Brown eased the fighter ahead. Like all Navy Corsairs, his plane bristled with weapons—including six .50 caliber machine guns, eight salvo rockets, and an egg-shaped bomb that contained napalm jelly.

Deckhands summoned Brown and nine other Corsairs into place. The whole aircraft carrier pulsed with the roar of the engines and the whine of the yellow-tipped propellers. In the *Leyte*'s tower, hundreds of well-wishers could be seen leaning against the rails, cheering. Many aboard this *Essex*-class carrier seemed to know that the squadron of Corsairs would be heading over the reservoir, in what promised to be a dangerous assortment of missions.

Brown gave a thumbs-up in recognition, then throttled his Corsair forward again. Behind him, serving as his wingman for the day's mission, was Lieutenant Thomas Hudner, piloting Corsair 205. Brown and Hudner had trained together at a naval air station in Rhode Island and had recently completed a lengthy tour of duty, cruising the Mediterranean from Crete to Monaco to Lebanon. Today they would stay in close radio contact and do their best to keep each other out of harm's way.

The pilots had run through their checklists, and the squadron was ready. At the signal, Brown raced down the long planked runway, gaining speed, slowly lifting the nose. At eighty miles an hour, he climbed off the deck and banked over the sea. Soon Hudner and the others were also launched into the sky, and the Corsairs fell into formation. Turning west, they sped over the shoreline and, two by two, vectored for the snowy mountains beyond.

≡

Jesse Brown and Tom Hudner hailed from dramatically different worlds, but they were the truest of friends.

Quiet, steady, cautious, Jesse Brown had a bashful smile offset by a sharp gaze that projected an air of purposefulness. Though he was only twenty-four years old, he had already flown twenty combat missions over Korea, and he was one of the most experienced pilots—and most reliable shots—on the *Leyte*. Devoutly religious, he often prayed in the cockpit before takeoff. He usually kept to himself while aboard ship. He neither smoked nor drank ("Make mine a gin and tonic—but hold the gin," he'd say when he found himself in a bar). He was smitten with his wife, Daisy, and their baby girl, Pam. He often spent his nights on the *Leyte* writing Daisy love letters or designing the dream house they hoped to start building back home in Hattiesburg, Mississippi, once the war was over. He had only a few more months of active duty left and then he could go home. The previous night, he'd stayed up late working on a letter to Daisy. He'd slipped it into the *Leyte*'s maildrop that morning.

Reserved and unassuming though he was, Brown was a celebrity in the world of aviation. He was the first black fighter pilot in the U.S. Navy, and the first African American to fly off an aircraft carrier.

He'd been photographed by *Life,* and he'd been the subject of newspaper articles all across the country. The other black men aboard the *Leyte*—stewards, petty officers, mechanics, deckhands—so revered Brown for breaking the color barrier that they'd recently pooled their money and surprised him with an expensive Rolex watch. Brown wore it with pride.

He'd come from humble beginnings. The son of black Mississippi sharecroppers, with Choctaw and Chickasaw blood also mingled in his veins, Jesse had grown up in a tin-roofed wooden shack without plumbing or electricity, in a little place called Lux, not far from Hattiesburg. The cabin was set beside some railroad tracks at the edge of the piney woods. "The whole structure sat on blocks, a pile under each corner," wrote Brown's biographer, Adam Makos. "When trains raced by at night, the cabin shook." Brown grew up in a world of mules and molasses, overalls and overtime. (During harvest season, the family liked to say they worked from "cain't see to cain't see.") Brown's parents, stalwart Baptists, grew their own vegetables and raised their own chickens, but they were forever in debt to the company store run by the landowner whose fields they worked. Mired in the monotony of their circumstances, the Browns urged their children to pursue an education.

Jesse took the lesson to heart. When he wasn't chopping cane or cotton, he usually had his nose in a book. A star athlete, he graduated as the salutatorian of his high school in Hattiesburg and won a scholarship to study architectural engineering at Ohio State University (which he chose because his hero, Jesse Owens, had gone there).

Brown's real passion, though, was aviation. When Jesse was six, his dad had taken him to a local air show that featured biplanes doing acrobatic stunts and wing walkers defying death, only to float safely to earth by parachute. Jesse was hooked: Ever since that day, he had wanted to fly planes. By lamplight, he read *Popular Aviation* magazine until the copies became tattered. People often scorned the naïveté of his ambition: "If Negroes can't ride in aeroplanes, they sure ain't gonna be flying one," a family friend had once assured him. But when President Truman, in 1947, began the process of fully integrating the United States armed forces, Jesse decided to enlist in the Navy to become a pilot. He trained for three hard years, in Illinois and Iowa, then in Pensacola, Florida, learning to fly Grumman Bearcats and Hellcats, then the Vought-designed F4U Corsair. He earned his commission as an ensign in 1949 and was assigned to Quonset Point Naval Air Station, in Rhode Island, where he met Tom Hudner.

Hudner was a winsome man with bright blue eyes and a lantern jaw. Twenty-six years old and a bachelor, "Hud," as he was known, liked to drink Scotch and smoke a pipe. He grew up in Fall River, Massachusetts, in a stately shingled Victorian house. His father had gone to Harvard and owned a successful chain of grocery stores. Even in the depths of the Great Depression, his family hardly suffered. Two of Hudner's brothers attended Harvard, and another went to Princeton. The family had a full-time maid, and young Tom often whiled away his afternoons playing lacrosse and running track. He was "movie star handsome," wrote naval historian and novelist Theodore Taylor, and spent his summers at the family's home on the coast, where he "did a lot of sailing and loved to read nautical books, especially about four-masters and Captain Horatio Hornblower." Hudner attended Phillips Andover, then the Naval Academy, in Annapolis, Maryland, where he graduated in 1947. For all the privileges of his upbringing, he had cultivated from his parents a commitment to duty and an unspoken sense of noblesse oblige. Hud liked the good life, but he wasn't a snob.

Unlike Brown, he'd never wanted to fly planes. His first sojourn in the skies was decidedly inauspicious—he was prone to airsickness, he found, and couldn't hold down his lunch. But as an Academy graduate, Hudner had to choose a path and put in his time for the Navy. His rationale for deciding to fly planes was superficial at best: Pilots were cool, he'd heard, and girls seemed to like them. All the same, Hudner grew to love flying, and he turned out to be good at it. He experienced a deep sense of calm in the air, and he seemed to have the right reflexes and instincts. He planned to make a career in the Navy. Everyone seemed to agree: Hud was a natural.

=

The squadron of Corsairs arrowed over the white gloom. Heading northwest, climbing from the coast, they loosely followed the road as it twined through the mountains, a road choked with scorched barricades and ruined vehicles. From up here in the relative warmth of their cockpits, moving at more than 250 miles an hour, the pilots couldn't imagine the travails that had taken place in this wintry outback. Brown and Hudner flew in silence, bouncing through the turbulence, occasionally snatching glimpses of each other as they flew side by side.

Forty-five minutes later, the planes came to a place where the country opened up to a valley that was studded with tents, girded with entrenchments, ringed with artillery, and marked by a fresh gash that proved to be an airstrip. It was a military tent city lofted in the mountains, a tabernacle of war: *Hagaru.* Smoke twisted from a dozen fires, and vehicles idled in white veils of exhaust. For Hudner, the frozen village brought to mind a painting he'd seen in his youth, depicting the winter encampment of a Roman legion dug in at the edge of Gaul.

Dozens of Corsairs and Skyraiders circled over Hagaru, awaiting orders from the dispatcher below. Brown and Hudner fell in with the flow of traffic and tried to follow the radio cross-chatter in their headphones. From their high perch, the circling pilots could see the transport planes parked on the rubbly runway, some disgorging reinforcements, others taking on casualties from the battlefield.

Finally, the pilots in the circling Corsairs got their orders: Head north toward Yudam-ni, then fly over the trackless barrens along the western shores of the lake. That morning, a large contingent of Chinese replacements, as many as ten thousand, had reportedly been seen marching south, aiming for Hagaru. The squadron's mission was to confirm the size and whereabouts of this enemy force—and, if possible, attack it.

Flipping on their gunsights, arming their salvo rockets, Brown and Hudner peeled off and headed north with the rest of the Corsairs. They dropped down low, almost to the level of the treetops, and followed a dirt road that skirted the reservoir. When Brown first flew over the Chosin country, a month earlier, he had found it a land of unsurpassed beauty—the luminescent blue lake surrounded by fall foliage and mountain meadows matted in the season's last wildflowers. He had written to Daisy about it. He'd never seen such a tranquil spot. But now it was covered with snow and scrims of fog, a gray and forbidding place.

The Corsairs searched for a half hour, covering a broad grid of terrain, but the recon mission was a bust: no sign of the Chinese. Wherever they were, they remained cleverly hidden. At three o'clock, about seventeen miles north of Hagaru, near a village called Somong-ni, an unseen Chinese gunman raised his weapon and fired into the sky. A bullet struck the thin aluminum skin of Corsair 211, Jesse Brown's plane, and pierced the oil line. It was a one-in-a-million shot. Brown didn't hear it over the vibrations of his engine.

The concerned voice of another Corsair pilot broke over the radio. "Jesse, something's wrong—looks like you're bleeding fuel!"

Brown, swiveling his neck, couldn't see anything. But Hudner was able to confirm it: "You've got a streamer, all right."

Brown checked his instrument panel and saw the needle on his oil pressure gauge dropping fast. The engine grew hotter and hotter as oil seeped from the line. Brown couldn't do anything to stop it. The gnashing was terrible. The friction was grinding away the innards of the plane. His engine was burning up.

He knew he couldn't make it back to Hagaru, and he was already too low to bail out and land by parachute. He had no alternative but to crash his plane—and do it in a hurry.

"I have to put it down," Brown radioed.

Scanning the topography, Brown couldn't find anything promising. It was the usual Chosin country—a confusion of steep ridges and ravines, much of it covered in runt pine trees. His engine's eighteen cylinders were misfiring, his propeller stammering and spitting smoke. Corsair 211 dropped into a steep descent. Hudner followed him, and they both hunted for a swath of level country. At ten o'clock ahead, a half mile toward the north, Brown spotted a relatively flat place, a bowl-shaped clearing framed by spines of rock. It seemed to be a mountain pasture where villagers took livestock to graze in the summer months. A boarded-up shack stood on the verge of the field. If he could glide his way over there, he might be able to lay his aircraft down.

"Taking departure," Brown transmitted as he zeroed in on the clearing. His voice was calm.

Hudner swooped beside him and talked him through the procedures for a crash landing. In short order, Brown jettisoned his belly tank, released his napalm, and fired his rockets into the hills—measures designed to lighten the aircraft and minimize the likelihood of an explosive fire.

As he neared his target, Brown saw that the pasture left much to be desired. Hardly level, it was crosshatched with small trees. Large rocks lurked beneath a veneer of snow. But it would have to do. Following protocol, Brown cranked his canopy backward until it clicked into place. The freezing air rushed over his face as the plane glided downward. Hudner was still at his wing, issuing encouragement over the radio.

Brown held the nose up as long as he could, but he was losing control: The plane had stopped flying. As Brown prepared to crash, Hudner pulled back on the stick of his Corsair and climbed to safety.

With a terrific shudder, Brown's six-ton aircraft struck the earth and sent up a cloud of snow and debris. The impact demolished the propeller blades, hurling fragments across the field. For a few seconds, Corsair 211 furrowed through the snow, groaning and buckling as it went. Finally it ground to a stop in a thin hiss of smoke.

My own dear sweet Angel, I'm so lonesome. But I try to restrain myself and think of the fun we're going to have when we do get together, so only a few tears escape now and then. The last few days we've been doing quite a bit of flying, trying to slow down the Chinese communists and give support to some Marines who were surrounded. Helping those poor guys on the ground, I think every pilot here would fly until he dropped in his tracks.

Don't be discouraged, Angel, believe in God and believe in Him with all your might and I know that things will work out all right.

We need Him now like never before. Have faith with me, darling, and He'll see us thru and we'll be together again before too long. I want you to keep that pretty little chin up, Angel, come on now way up. Your husband loves his wife with all his heart and soul—no man ever loved a woman more.

40

THE BRIDGE OF LONG LIFE

Hamhung

That same week, Lee Bae-suk touched down in his native province after four long years away. His assignment had finally come through: He was going to serve as a translator, as well as a military policeman, for the X Corps forces in Hamhung. His plane, a four-engine C-54, rolled to a stop at Yonpo Airfield, and Lee, churning with emotions, pressed his face against the window. Were his parents alive? His brothers and sisters, his cousins and friends? Would his neighborhood still be standing? What would happen to Hamhung when the Americans left? Lee could scarcely work through the welter of feelings that coursed through him. He yearned to see his home but feared he wouldn't recognize it. He was worried that he had arrived too late.

Lee and other two translators stepped off the plane into a blustery day, sunny but bitterly cold. Enormous piles of crated supplies sat on the Yonpo runway. A squadron of Corsairs stood idling, preparing to take off for the highlands around Chosin. An officer from X Corps greeted the three translators and ushered them into the base. They were driven to Hamhung's city hall, which had been taken over by the American military. Lee was told to report to the Criminal Investigations Division, down the hall, to undergo questioning. Lee's gut clenched at this: He still had not told anyone about his past—that he had grown up in North Korea, in this very city. He feared that if he did, he would be arrested as an infiltrator and a spy. He was certain that the investigators would see him as the enemy and send him to jail.

An American officer led off with a few questions. Lee fibbed, saying he had been born and raised in Seoul. His northern dialect might
have given him away, but luckily, the Korean officer on duty had been
pulled aside on another matter. Lee sailed through the questioning,
and it was official: He would serve in a unit of the X Corps military
police, attached to the U.S. Army's Third Infantry Division. He was
given a uniform and a rifle, offered C rations to eat, and told to report
back in the morning for duty. They told him he would be working at
the most important checkpoint in the city, guarding a large bridge
that spanned the Songchon River. It was called Mansekyo—the
Bridge of Long Life.

=

Once again, Lee could not believe his luck. The bridge was a
short walk from his family's house. Arbitrary forces of Army manage-
ment, abstract orders from an unthinking bureaucracy, were directing
him straight to his neighborhood, to the place where he wanted to
be. He remembered with fondness the many festive times, at New
Year's, he had crossed over the bridge with his brothers and sisters, his
mother and father, his cousins and school chums. Mansekyo lay in the
heart of the city. It was a place he held dear—and, if true to its name,
it was a place of good fortune.

The next day, Lee was driven by jeep through the city. He could
see that Hamhung was in shambles, its neighborhoods choked with
rubble, its shops and markets empty, its citizens buffeted, weary, and
tense. Many people milled in the streets, their belongings bundled on
their backs or piled high on sidewalks. No one seemed to know where
to go, or what the future would bring. American soldiers moved
through the streets on crisp errands. Machine guns peeked out from
behind barricades of sandbags. At a roadblock, MPs waved a truck-
load of captured Chinese soldiers through—they were bound for X
Corps headquarters, to await interrogation.

Adding to the confusion in the city were the refugees from the
countryside. Many tens of thousands of people had streamed down
from the mountains—some to elude the violence of the Chosin fight-

ing, some to find food, still others harboring the more ambitious hope of boarding ships bound for another life in the South. He could hear their cries in the streets.

Lee hardly recognized the city, parts of which the Americans had strafed and bombed several months earlier as the North Korean troops pulled out. Now it was the Americans who would be pulling out, and nearly everyone in Hamhung could feel a riptide of calamity, an awareness that the city would soon be engulfed in sorrows impossible to measure.

Lee could hear artillery, coming from somewhere off in the reservoir country. The departing Americans had plans to bomb the city, to destroy all factories, to blow all bridges, to leave nothing behind of strategic value. Hungnam and Hamhung were doomed sister cities, and everyone sensed it. Placards would be stapled around town and leaflets dropped from the air, warning civilians in certain areas to begin evacuating their homes. The American exit plan would deny the Chinese their spoils—but it was the civilians who would suffer most.

In another week or two, the Americans would probably be gone. And when they left, who could say what would happen? As the Chinese and North Korean armies, addled and starved, flooded into what was left of the city, the civilians expected to see spasms of predation, crimes of venality and vengeance. Now, on the streets, the sense of unease was palpable, like a rank fog hanging over every nook of the city, over every person and every house.

The jeep approached the bridge, and Lee stepped out. A military checkpoint had been set up on the east side—the *city* side—of the bridge, with a kiosk, a machine gun nest, and a tank parked nearby. This, he was told, would be his post for the next several weeks. He grabbed his rifle and took his place, as a sentry and an interpreter, on the Bridge of Long Life.

41

DOWN TO EARTH

North of the Reservoir

Lieutenant Tom Hudner circled the crash site, despairing for his friend. The plume of snow had cleared, revealing the full extent of the wreckage. The long nose of Jesse Brown's Corsair was badly misshapen—it had contorted with such force that it was nearly cracked in half. The plane was a tangle of ruptured hoses, ripped wires, and shorn cowling pieces.

Though Hudner repeatedly tried to reach Jesse over the radio, he got only static. But when he circled again and came in low, he took heart: Brown was waving up at him.

At that, Hudner heard voices cheering over the radio. He had momentarily forgotten that, higher up, other Corsairs from the squadron were waiting in a holding pattern. They'd been watching the disaster unfold. "I'm calling for a helicopter," one of the pilots broke in. "And Tom," he added, "once Jesse gets clear, destroy the plane." Hudner understood. Better to ruin your aircraft—blow it to smithereens if you have to—than let all that expensive Cold War technology fall into enemy hands.

Then the radio waves crackled: *"Mayday! Mayday!"*

Brown was still inside his cockpit. Was he hurt? Why didn't he close his canopy? It was freezing down there.

Now Hudner spotted something more alarming: Smoke was curling from Brown's plane. The wreck had caught fire—it was probably a magnesium blaze, coming from the engine itself. Hudner knew that just in front of Brown, inside the aircraft's twisted nose, was a two-hundred-gallon gas tank. If it ignited, there'd be no hope.

Brown kept waving.

Hudner orbited again, then once more. He knew what he was supposed to do in this situation. He'd studied it in the manuals. He was supposed to steel himself and turn away. But a worm was turning in his conscience. He was thinking of taking another course, something that went against Navy rules and might get him court-martialed.

He banked his Corsair and dove toward the pasture. "I'm going in" was all he said over the radio to his squadron mates. He wasn't asking for permission or advice. It was a statement. Up above, no one uttered a word.

$$\equiv$$

Tom Hudner aimed straight for the hillside, toward a promising spot near the downed plane. Facing none of Brown's mechanical challenges, Hudner was better able to control his descent and feather his landing. Even so, he crashed with terrific force, in a tumult of wrenched metal and snapped propeller blades. His windshield shattered. Once he'd skidded to a halt, Hudner slid back the canopy and tried to crawl from the cockpit, only to find himself wincing in pain. He thought he might have cracked a vertebra. His engine steamed in the cold air. He turned and saw that Brown's plane was still smoldering.

Hudner knew he had to act quickly. Somehow he lifted himself out of the cockpit and hopped off the wing into the snow. Grimacing through the pain, he moved up the slope toward Brown's Corsair. But then his heart skipped a beat: Tracks, apparently fresh ones, were stitched across the snow. Were they Chinese? North Korean? The shack at the edge of the field appeared to be abandoned, but maybe someone was hiding inside? The mountain winds roared in his helmet. He thought he heard voices. He pulled out his .38 revolver and fired a shot into the air. Then another. He waited a few moments, but no one stirred from the shack.

Encouraged, Hudner resumed his painful climb across the hill. He approached the right side of Brown's plane and clambered up onto the wing. As he looked down at Brown, Hudner's spirits sank. According to biographer Adam Makos, Brown's lips "were blue and his ears looked frozen and brittle. He was shivering wildly, his arms folded, his breath puffing. The cold had curled his fingers into claws."

"Jesse."

"Tom," Brown said through chattering teeth. "I'm pinned in here."

"Helicopter's on the way," Hudner said. He looked at his friend. Brown's lower body had twisted in the rending metal. One wall of the cockpit was stove in. Hudner feared that Brown might have broken his back. He might have had internal injuries, too, and he could be in shock. His face was waxy. Brown's ears ached, and his fingers were so stiff he could hardly move them. Hudner pulled a knit hat over Brown's head and wrapped his benumbed hands in a scarf.

Appraising the cockpit, Hudner was able to piece together what had happened in the moments immediately after the crash. Brown, in his scramble to get out of his parachute and escape the smoking plane, had hastily removed his gloves and helmet and dropped them, just out of reach, on the floor of the cockpit—only to realize that he was trapped. (The helmet contained the microphone and headset, which explained why he'd been unable to communicate with Hudner by radio after the crash.)

The smoke was growing thicker, and Hudner was worried that the plane might explode. Brown was, too. "We gotta do something to get out of here," Brown said.

"Let's see what's got you pinned," Hudner said. He leaned into the cockpit and batted away the smoke long enough to determine the problem: An extension of the instrument panel was crushing Brown's right knee. He was jammed in tight, but Hudner had to try to pull him out. He found a good foothold, then grasped Brown under his arms. On the count of three, he yanked with all his might, but nothing budged. Hudner revised his hold and heaved again. It was no use: His friend was truly stuck. It was going to take a crowbar and an ax, Hudner thought—maybe even a cutting torch—to extricate him.

Brown remained stoic, but he was suffering, his energy ebbing, his breathing strained. Hudner eyed the billowing smoke with alarm. He hopped off the wing and scooped some snow, shoving it into the engine compartment, where the smoke was thickest. He packed in more and more. The fire smothered a bit but then roused again.

At about 3:40 p.m., Hudner hurried over to his own plane and got on the radio. Several Corsairs from the squadron were still circling overhead, their pilots scouring the surrounding country for any Chinese who might be drawn to the crashed plane. "Jesse's pinned inside," Hudner transmitted. "His back may be broken. But he's got all the heart in the world."

Hudner peered into the skies and noticed that the Corsairs were circling in a clockwise pattern—a Navy signal that meant that a rescue helicopter was en route. This was especially welcome news: With daylight starting to drain from the sky, Hudner knew the planes would have to make their exit soon and return to the deck of the *Leyte*.

Around four o'clock, the Corsairs dove and, one by one, buzzed the pasture. It was their way of saying farewell. Brown raised an arm in reply. They wagged their wings as they passed over the two downed planes, then headed east over the reservoir.

They were alone now, two injured aviators in enemy territory, surrounded by frozen wasteland. Although, Hudner feared, maybe they weren't alone. He knew it would be only a matter of time before the Chinese found them. The trail of smoke from Brown's plane had announced their presence. Hudner's thoughts spun. Would the Chinese kill them on the spot? Or take them prisoner? What would captivity be like in such a severe country? Shivering in the slanting light, he fumbled for his .38, but he recognized what feeble protection it afforded.

Hudner turned to his friend again. He could see that Brown was fading. His breathing was shallow and labored. His head drooped. His eyes had become vacant and lusterless. After a few minutes, Brown broke the silence: "Tell Daisy how much I love her."

"I will, Jesse," Hudner said.

≡

The Sikorsky looked like a dragonfly, buzzing awkwardly over the mountains, the throb of its rotors echoing off the rocks. Its pilot, First Lieutenant Charlie Ward, fought the winds as he tried to land the bulbous-snouted creature. Hudner activated a red smoke flare to help guide the pilot in. Ward plopped the chopper down, its three tires sagging in the snow. He kept the Sikorsky running; he feared that if he killed the engine, he might not get it started again.

Ward hopped out into the snowy tempest stirred up by the blades and greeted Hudner in astonishment. The two men were friends. Hudner tried to explain the situation, shouting over the whir of the chopper. Ward squinted and said, "That Jesse over there? Aww, shit."

Ward produced an ax and a fire extinguisher from the Sikorsky, and he and Hudner hastened to Brown's plane. Hopping onto the wing, Hudner gripped the ax first and smashed it into the buckled instrument panel that was pinning Brown in place. But the blade only bounced off. He tried again and again, adjusting the angle. He used the handle as a lever to pry back the encumbering bulge of metal, but to no avail.

Then Ward, who'd been working over the engine with the fire extinguisher, took a turn with the ax. "Hey Mississippi," Ward said. "Hang in there." With all his strength, Ward wailed away. Searching for leverage anywhere he could, he tried to pry his friend loose. Brown, according to one account, "half smiled as he watched, glassy-eyed, the feverish efforts." But nothing was working. He was still trapped.

And he was slipping fast. The cold was taking him. Ward and Hudner both could see it. They considered the possibility of amputating Brown's leg to free him. Hudner had a sharp knife strapped at his hip, but he couldn't go through with the abhorrent task.

Brown lost consciousness. His breathing was barely detectable, just thin wisps of air. Then he stopped breathing altogether. Ward laid down the ax. "He ain't moving," he said.

Hudner was distraught, pacing beside the plane, his mind in a fugue state. "They must have a cutting torch at Hagaru!" he shouted.

Ward inspected Brown more closely and shook his head. "Think he's gone, Tom." They both examined him, and there was little doubt: Ensign Jesse Brown was unresponsive.

Hudner couldn't bring himself to accept this uncomfortable truth, nor to confront the one that immediately followed: They had to leave now, and abandon Brown here on the hillside, perhaps to be stripped and desecrated by the enemy. The Sikorsky wasn't equipped with night navigation, and they had only a few minutes before darkness would descend. They thought about hacking off Brown's pinned leg so they could at least remove the body. But in the end, they thought better of it. The magnesium fire continued to smolder and could ignite at any moment. They would try to retrieve Brown's remains in the morning.

Still, Hudner was in a muddle. "You coming or staying?" Ward asked. He reminded Hudner that if he remained here on this mountain, he would surely die.

Coming to his senses, Hudner followed Ward back to the idling Sikorsky, and they crawled inside. As the helicopter rose above the wreckage, Hudner pressed his face against the glass nose and stared. He had destroyed an expensive plane. He had violated Navy rules. His friend was dead and left to be scavenged. He himself was injured. And in a few days, he felt sure he would face a court-martial.

The chopper bobbled over the darkening peaks and then, gathering momentum, sped south toward Koto-ri.

> *If only my heart could talk—if only my hungry arms could enfold you. . . . Often at nite all the loneliness of the day seems to descend upon me. Then all the tears that I've been holding back refuse to be held any longer, and I just lay there in loneliness and misery and cry my heart out.*
>
> *Darling, I'm going to close now and climb in the rack. I have to fly tomorrow. But so far as that goes my heart hasn't been down to earth since the first time you kissed me, and when you love me you send it clear clear-out-of-this-world. I'll write again as soon as I can.*
>
> *Your devoted husband, lovingly and completely yours forever,*
> *Jesse*

42

THE MOST HARROWING HOUR

Koto-ri

At nine thirty on the morning of December 7, the sound of propellers cut through the air. Three Flying Boxcars thrummed across the sky and circled over the Marine encampment at Koto-ri. The twin-tailed, fat-bellied transports seemed to strain under the weight of their loads. Ground signalers scattered across a frozen field just outside the hamlet and began to unfurl bright orange panels to mark the borders of the drop zone. Lieutenant Colonel Partridge watched through a pair of binoculars; General Smith, who had moved his headquarters here from Hagaru, stood at his side. Hundreds, possibly thousands, of Marines watched as well, knowing their fate hinged on what the skies today might yield.

One of the Boxcars dropped to about a thousand feet over the ground. From the surrounding hills, enemy fire crackled at the slow aircraft. The plane's rear bay doors cranked open. In its belly, crewmen untethered all but one of the ropes that cinched the treadway piece to the ribs of the plane. The pilot, spotting the orange panels below, descended to eight hundred feet, feathered back on the throttle, then gave his command to the cargo master: *Drop!*

The procedure that had been improvised for discharging the 2,900-pound bridge section was elaborate, a Rube Goldberg progression of interconnected tasks that had to be flawlessly executed in a few seconds. On the pilot's cue, a crewman standing deep in the bay would chop the last remaining rope with an ax while another man stood by to smack the bridge section with a sledgehammer—a final nudge to eject the big piece of steel from the plane's abdomen. Simultaneously, a third crewman was supposed to trip a spring-loaded drag parachute designed to guide the piece safely out into the slipstream as the plane pulled away. This smaller chute, in turn, would activate the two larger chutes, hopefully breaking the treadway piece's fall.

All of these maneuvers had to happen with clockwork precision, but the pilot also had to be flying at just the right altitude, and at just the right speed, if he hoped to hit the drop zone with any accuracy. The target was a confined area, a few hundred square yards, in the flats hugging the town. If the treadway piece landed directly on Koto-ri, crowded as it was with installations, warming tents, and medical wards, it could cause a great deal of damage, maybe even kill someone. If the bridge section drifted outside the perimeter, the Chinese would claim it. And if it glided off into the surrounding mountains, it would be impossible to retrieve.

Now the first plane exhaled its cargo. The Boxcar bumbled in the air, in sudden reaction to its lightened load, but then stabilized. In rapid succession, the freight's parachutes opened as they were supposed to, and it slowly sailed toward earth. The girder landed, unharmed, in the field, not far from one of the Brockway utility trucks. The Marines, watching from their stations in Koto-ri, erupted in war whoops and clapped their mittened hands.

The other two planes swooped down and dropped their treadway segments as well—both successfully. Then five more Flying Boxcars arrived, bearing five more bridge sections. One of the pieces drifted outside the perimeter and, as predicted, the Chinese fell upon it and hauled it away. This was doubly worrisome to Partridge: The enemy, upon inspecting the fallen bridge section, would fully apprehend, if he hadn't already, the American intention to rebuild the span at Funchilin Pass. The Chinese would only redouble their defenses at the bridge site.

Then one of the treadway sections dropped too quickly—one of its chutes didn't open properly. It was damaged beyond repair when it smashed to the frozen ground. But three more sections landed safely. By afternoon's end, the Brockway crews were able to recover six intact pieces, and that was enough. Meanwhile, another plane arrived to drop a substantial supply of precut plywood to serve as flooring for the bridge. Partridge was relieved: No planes had crashed, and no one had been hurt. He had everything he needed to carry out his plan. His engineers could get started.

≡

At midmorning the next day, the two huge Brockway trucks, loaded with equipment and bridge parts, clanked away from Koto-ri in a blizzard. Partridge and his Marine engineers were accompanied by members of the Army's Fifty-eighth Treadway Bridge Company. Platoons from Homer Litzenberg's Seventh Marines worked out ahead of the two big trucks, clearing the country of the enemy as they went. The fighting was fierce, all the more so because the Marines had no air cover—Corsairs couldn't fly in the whiteout conditions of this snowstorm. Each Brockway groaned down the road like a queen bee, fat and pampered, zealously protected by her own hive of soldiers.

The convoy made only halting progress toward the blown bridge. The Chinese kept dropping shell fire on the road, trying to destroy the two big trucks and their precious cargo. With darkness descending, Litzenberg ordered the Brockways to the rear of the column. The mortar barrages were too intense, and he couldn't risk damage to the trucks. Not only were the Brockways carrying the vital bridge parts, but they were also rigged with the hoists, cranks, winches, and cranes necessary for the span's construction. Said Partridge: "The trucks, with their hydraulic operating systems, were vulnerable items of essential equipment; we had to get them out of there."

So the Brockways were turned around and driven to a protected place, tucked back from the road. The first truck nosed onto what appeared to be a good, flat parking space. But then its driver heard a frightful cracking sound and realized, too late, what was happening. This spot was not solid ground; it was a frozen-over pond. Pockets of spring water burbled beneath the ice, and then it collapsed. The Brockway, wallowing at first, sank up to its radiator in frigid water. Its engine wheezed and died. The driver couldn't get it restarted—the motor was drowned. He was truly stuck.

The other Brockway was brought over to the pond's edge to yank her sister out. For a half hour, the engine revved and chuffed. Spinning wheels sang hot on the snow, flinging clods of dirt. Finally, the disabled truck emerged from the pond, rimed in ice, her engine apparently ruined. Partridge and his men fell into despair. "They were sick," one Marine account put it, and they stood hunched in disbelief, "watching through slitted eyes." As night fell, Partridge tried to reckon with the situation. "For me," he said, "this was the most harrowing hour of the campaign."

43

THE CROSSING

Funchilin Pass

The blizzard that had blown all day continued well into the night. But sometime in the early morning of December 9, the Marines at Koto-ri caught a hint of hope: They observed a single star winking through gauzy openings in the clouds. Many, taking it as an auspicious sign, came to call it the "Star of Koto-ri." Said Staff Sergeant Manert Kennedy: "Seeing that star was like being given another shot at life. We had no idea if we were going to make it out of there. The star was a godsend, it was real—it meant the world to us."

Sure enough, by first light the snowstorm had cleared, and this happy turn in the weather changed everything. Now the Corsairs could return to the skies to menace and scatter the Chinese along the road. The engineers, with their one functioning and now grossly overloaded Brockway truck, could expect to reach the bridge site unmolested. The thousands of Americans trapped above the blown span could sense their salvation now.

All morning, the Marines waged a mean battle at the site of the bridge. The mountains shuddered with artillery, and the Corsairs reddened the hillsides with their rockets and canisters of napalm. The artillerymen often employed "proximity fuze" shells—special munitions equipped with sensors that bounced radio waves off the ground as they hurtled through the sky; the shells were set to detonate at a predetermined height above their target, blasting thousands of metal shards down upon entire hillsides of the Chinese. By midday the enemy fortifications had crumbled. More than a hundred Chinese soldiers had surrendered. The word was conveyed by radio, and the engineers, guarded by a weapons company, hastened the three miles from Koto-ri down the MSR, reaching the site without incident. Partridge's men, along with those on Lieutenant Ward's bridge-building team, went to work.

But within a few minutes, Partridge and Ward made an awful discovery: Over the past few days, the Chinese had managed to destroy the bridge further. The gap was now twenty-nine feet across, some nine feet wider than Partridge had estimated. This posed a problem, for the treadway sections, bolted together, reached only twenty-two feet. The engineers scratched their heads, did some more measuring, and admitted that they were stumped.

Luckily, one of the enlisted engineers had spotted a pile of railroad timbers nearby and proposed that carpenters could build up a "crib" of interlacing lumber on a ledge that still clung to the bridge's undergirding, about eight feet below the road level. This shelf was just wide enough to make up the difference between the treadway span and the actual gap.

Partridge gave the idea his approval. More than a hundred men were thrown into the project—including all the able-bodied Chinese prisoners the Marines could commandeer. Shuffling on frostbitten feet, they fetched railroad ties, they assembled a foundation of sandbags, they hammered and sawed, they laid piece upon piece. Slowly, a scaffolding of wood rose from the depths until it reached the level of the road. But this hollow abutment of rough-cut lumber was dangerously wobbly; Partridge needed ballast to stabilize it. The engineers, having run out of sandbags, tried to gather sufficient quantities of dirt, yet the frozen earth proved impossible to excavate. Then a solution, grotesque but perhaps inevitable, took shape: They would fill in the gaps with human bodies.

The battle for the bridge site had produced hundreds, possibly thousands, of Chinese corpses. They were strewn on the bloody hills, lying in ditches and draws, wedged into foxholes. They seemed to leer down at the pass they had given their lives to defend. If using them as a structural material seemed a ghastly expedient, it was perhaps no worse than leaving them where they were, out in the snow squalls, frozen in twisted shapes, their blanched faces bearing expressions of anguish and pain. At least some of them would be given a kind of burial, within the substructure of the bridge.

A work detail began the grim task of collecting bodies and dropping them into the span's interstices. It seemed strange but not so terrible at the time. It was another horror to add to those they'd seen and done, and had done to them—another bruise to the soul. The lattice crib was filled with human ballast, and the engineers declared the structure stable enough to accept the full weight of the bridge pieces. The lone Brockway truck was brought forward, and its hydraulic crane swung the steel segments into place. Workmen crawled out to bolt together the spacers and fasteners—performing "high-wire acrobatics," according to one Marine account, "while from the near hills Chinese snipers added to the entertainment, trying to pick off individual engineers swinging out over the void." As the construction proceeded, Marines patrolled the hills, dislodging the last nests of the Chinese.

The engineers worked for several hours. Finally, crews laid a course of thick plywood sheets between the metal treadways, and the span was complete. Partridge ordered a test run. A jeep eased out over the chasm. The bridge settled and creaked, but it held.

Partridge couldn't contain his glee as he drove to the top of the pass to alert the waiting train of division vehicles that the span was repaired and the breakout could begin at last. Stretching to Koto-ri and back toward Hagaru, a convoy numbering more than a thousand vehicles had been waiting for this moment, fighting off the Chinese all the while. It was a snake of rolling machinery, bumper to bumper, ten miles long: ambulances, transport trucks, half-tracks, artillery pieces, tractors, bulldozers, snowplows, tanks. Some fourteen thousand Marines and Army soldiers, as well as Royal Marines, would be marching or motoring down the mountain. They had come to the reservoir as a fighting force, and they would come out the same way, with their guns and equipment intact.

Partridge returned to the bridge site and, in the lengthening shadows of the late afternoon, waved the first dozen vehicles across. The column seemed to be moving smartly along when a large bulldozer crossing the span snapped through one of the plywood floorboards. The piece of equipment was too heavy—the sheet of wood gave way with a crack, and the nose of the bulldozer tipped into the abyss. This was too much for the terrified driver, who slid off and slunk back to the north side of the bridge. The whole train of vehicles was stopped, and would remain so until this ponderous obstacle could be removed. Partridge mulled over a solution. Darkness had fallen, making the task before him even more daunting.

Then a technical sergeant named Wilfred Prosser pressed forward and volunteered to rescue the stricken piece of machinery. By most accounts, Prosser was the best dozer operator in the whole division and was not wanting in courage. He sidestepped onto the bridge and climbed to the seat of the tipped bulldozer. By deftly working the blade, Prosser was able to right the vehicle. Then, guided by flashlights, he eased it backward, drawing applause as he safely returned to terra firma.

Partridge's engineers had to find a way to repair the cracked bridge. They decided they should yank out the plywood center boards and respace the steel treadways a few inches farther apart. In this new configuration, wider vehicles would ride along the outer lip of the span while narrower vehicles would ride the inside lip. This should work, Partridge reasoned, but it meant the drivers would have precious few inches to spare. With the precut plywood sheets removed, the gap between the parallel metal spans would be nothing but black air. The drivers would have to hold their tires in exactly the same position for the crossing or risk a drop into the void. Each vehicle would have to inch along, with teams of observers holding flashlights front and rear.

Partridge's men removed the plywood boards and began to respace the steel treadways. In another hour it was done, and the vehicles were moving again. Their drivers fought off vertigo as they crept across the gulf. They had to trust their pedestrian guides—and not look down. Partridge's modified bridge was treacherous, but it seemed to be working.

=

Eight hundred yards down the mountain, however, another obstacle loomed: The blown railway trestle still lay across the road, like an unwanted toy tossed aside by a surly giant. To some of Partridge's men, this seemed an even tougher impediment than the blown bridge. The sappers were going to have to blast apart the metal monstrosity, piece by piece. It could take hours—and Partridge didn't have hours.

One of his lieutenants, David Peppin, went to inspect the trestle with a small team and a bulldozer. Peppin observed that there was a stream issuing from the mountain nearby. It was mostly frozen, but it still seeped and trickled. Where it met the road, the water fanned out in frozen sheets beneath the railroad trestle.

Peppin had an idea. "Just for the hell of it," he recalled, "I told the bulldozer operator to butt his blade up against the trestle and see what would happen." The driver figured it was a quixotic move, but he was willing to try. When he did, the result was almost comical: He smacked it once and the whole piece, many tons in weight, slid effortlessly along like a puck on a rink. Within a few seconds, the little dozer had pushed the trestle off the road.

Nothing stood in Partridge's way. The road to Hamhung and on to the sea was clear. Smith's men, who had been steadily flooding into Koto-ri over the past several days, were finally free to break out. Now the long file of troops and vehicles began to spill from the mountains and cross over Partridge's unlikely construction—a bridge delivered by parachutes, assembled by acrobats, and buttressed, in part, by the enemy's frozen flesh. "I'll get you a goddamn bridge," Partridge had said—and he'd made good on his promise. In a few hours, once everyone had passed over it, his men would blow it up again.

Through the night, the men and the machines kept coming, easing down the steeps toward Hamhung. The procession was quiet, save for the distant rumbling of artillery. People hardly spoke. They sensed that their ordeal might be over, and with that realization came flood tides of emotion. On the road that night, Partridge found the mood hopeful, but also spooky. "There seemed to be a glow over everything," he recalled. "There was no illumination and yet you seemed to see quite well; there was the crunching of many feet and many vehicles on the crisp snow. There were many North Korean refugees on one side of the column and Marines walking on the other side. Every once in a while, there would be a baby wailing. There were cattle on the road. Everything added to the general sensation of relief, or expected relief, and was about as eerie as anything I've ever experienced in my life."

Manert Kennedy thought the cries of the refugees had a distinct ebb and flow, like the paroxysms of a wounded creature. "Those poor folks out there," he said. "A collective moan came out of them, it had a rhythm to it. It was an ungodly sound of hopelessness, of helplessness. It chilled me right down to the bone. I can still hear it."

44

WE WILL SEE YOU AGAIN IN THE SOUTH

Hamhung

Lee Bae-suk found it surreal to be standing all day with a rifle slung over his arm at the Bridge of Long Life, a mere handful of blocks from the place of his birth. He felt like an interloper, a ghost in his own home. His job was to interrogate people crossing over the bridge, to check their passes and keep an eye out for suspicious activity. He scanned the crowds, looking for faces he knew but dreading what would happen if a face he knew spotted him. Lee couldn't tell his American comrades who he really was, and he hated the dissembling this entailed. The Americans, he'd heard, believed that all North Koreans were Communists. He tried to feign disinterest in his surroundings, as though this were just any assignment, in any place. He had a job to do, and he had to work hard to assume a stern and businesslike demeanor.

Several days went by. The work at the bridge was unceasing. Lee never got a break. He had hoped that he might find a lull, a free hour or two to peel away from the checkpoint and disappear into the city to hunt down his family. He tried to work up the nerve to ask his American superiors for an afternoon off. But he dared not—not yet, anyway. Such a request might rankle them, he feared, and it might raise suspicions.

Then, one morning, General Almond and the American command issued a new order: At noon that day, Hamhung was to be locked down. No one would be allowed to cross the bridge and enter the city. Except for American or South Korean soldiers holding the appropriate pass, all traffic must cease—indefinitely.

The purpose of the order was to relieve congestion in the over-crowded city so that the Americans could organize themselves and get smoothly on their way to Hungnam. It was designed primarily to keep out the flood of civilian refugees from the mountains, and any North Korean guerrillas or Chinese infiltrators who might be hiding, in various disguises, in their midst.

But the ban applied equally to locals: Any Hamhung citizens who, in their random comings and goings, happened to be outside the city when the order was issued could not pass back inside to resume their lives and reunite with their families. At the stroke of noon, the gates would shut. These hapless people, many thousands of them, would be trapped outside the city until the Americans left and the moratorium was lifted. Almond's order was unequivocal—no exceptions.

A loudspeaker was mounted on a jeep beside the checkpoint. At noon, Lee was told to announce the order. He hated to do it, but he had no choice. When his words issued from the squawky speaker, a wail of panic and outrage arose from across the river. Crowds began to teem on the other side of the bridge. Lee could hear their wild pleas. People were stunned. They cursed at the authorities and shook their fists in the air. Some dove into the freezing cold river and tried to swim across. Rumors circulated that the Americans had placed explosives in the undergirding of the bridge and were going to blow it to prevent any more crossings.

Lee had never seen such a heartbreaking sight: starving refugees from the mountains, wizened grandparents, infants crying at their mothers' breasts. Farmers with pots tied around them, clanging mis-cellanies of vessels stuffed with their last belongings. These people had come down from the Chosin battlefields in a great hegira, had braved the snows, had risked their lives—only to arrive at this implau-sible terminus at the moment the gate had snapped shut.

And Lee was there to stop them, to dash their hopes and turn them around. What an awful assignment it was. His orders were to halt them at the west side of the bridge and, through the tinny ampli-

fier, command them in the boldest possible language to walk no further. If they kept coming, Lee was told to point his rifle in the air and
fire a warning shot. And if they still kept coming, he was to shoot
directly into the crowds.

This, he felt, he could not do. He could not fire on his own people. He prayed it would never come to that.

$$\equiv$$

A few days later, Lee, posted at the bridge, worked up his courage
and asked his American superior if he could take an hour's break. To
his surprise, the officer said, "Sure, go ahead." Lee thanked him and
vanished into the crowded streets. He knew he had to make haste.
The layout of Hamhung's inner city was hardwired in his mind—he
knew every alley, every shortcut. His army boots clomped on the
pavement as he burrowed into the old neighborhoods and worked his
way toward the compound where his family and his uncle's family had
lived. With every footstep, his excitement grew. A left turn, a right,
and then he arrived. He exhaled in relief: The house was still standing. It had not changed much in four years, but it seemed quiet and
darkened inside, as though it had been abandoned.

He approached the door with apprehension and knocked. A few
moments passed, but no one answered. He knocked again. "Papa!"
he yelled. "Mama! It's me—Bae-suk!" When no one answered, he
opened the door himself. After all, it was his house.

He ducked his head inside and peeked into the shadowy anteroom. The place seemed in disarray, furniture stacked up, belongings
scattered about. He peered deeper into the house and heard a gasp.

"Bae-suk! Is that *you*?"

All at once, his family came rushing toward him—his mother and
father, his brothers and sisters. They smothered him with hugs and
kisses. His father stepped back and regarded him in astonishment:
Who was this grown man, in an army uniform, wearing a helmet,
looking so responsible, so serious? Tears welled in his parents' eyes,
but they hid most of their feelings inside. They were stoic people, not
given to displays of emotion. But Lee could tell something was wrong.

Worry lined their faces, worry that hung like a sickness in the room, blackening everything, canceling whatever joy his parents felt at his surprise homecoming.

Then Lee realized that his baby sister was missing. He looked around the house. "Where is Sun-ja?" he asked. She had been only three when he left for the South. She would be seven now.

His father shook his head and looked away. By the mood that fell over the room, Lee decided she must have died.

"No, no," his father said. "Sun-ja is fine. But we are afraid we'll never see her again."

"Why, Papa? What is wrong?"

His father gestured at the stuff stacked around the house. "We are taking passage to the South. We're leaving tomorrow. All the arrangements have been made." He explained that a truck would be taking them to Hungnam, where a ship awaited. They were leaving this country for good. They would take a few things with them, but most of their belongings would be dispersed to friends and distant relatives. They were going to make a new life for themselves in the South. "We had planned to go to Seoul," his father said, "in hopes of reuniting with you."

"And Sun-ja?"

"She is on the other side of the river." Lee's father explained that she had been visiting their cousins when the Americans issued the order: No one was to cross back into the city. The bridges were closed. He had beseeched the police, he had tried to explain the situation to the Americans, but to no avail. Sun-ja was trapped on the other side, and they could not reach her. They faced a predicament: Would they leave behind their youngest daughter forever? Or would they scrap their plan for escape, stay at home, and take their chances in a city that was about to be convulsed by unimaginable violence and possibly destroyed?

Lee had heard all he needed to hear. "I must go," he said. "I will try to find Sun-ja." He had been at the house no more than five minutes. He turned and stomped out the door.

=

As Lee trotted back toward the bridge, a strange feeling swirled inside him. Why was he here? It was as though fate had guided him to this precise place, at this precise moment. He thought about the situation and decided there was no other way to understand it. He was a humble and earnest young man, but now he was impressed with a sense of mission: Only he could save his sister; only he could make his family whole. Someone who knew this city, someone who spoke English, someone connected to the U.S. Army, someone the authorities could trust—and someone assigned to guard the very checkpoint his sister could not pass through.

He knew he had to risk everything. He returned to the bridge and went straight to his American supervisor. "There is something I must tell you," Lee said. "I was not born in Seoul. *This* is my home. This is where I was raised. My family is here. They are leaving tomorrow on a ship. But my little sister is over there." He pointed across the river. "She cannot get across."

He wasn't sure the American understood him. He was still uncertain how proficient his English was. He feared he'd said too much, and his fear was legitimate: Here in Hamhung, over the past week, Lee had seen young North Korean men, their covers blown, their identities revealed, brutally beaten and dragged off to jail.

"My family," Lee blurted out. "They are not Communists. They have never been Communists. They want to live in the South."

The American officer stood in silence for a few long moments, processing what he'd heard. Then he nodded and said, matter-of-factly, "Okay." He motioned toward a nearby jeep. "Let's go."

He cranked the engine, and they slipped through the checkpoint barricades and sped across the Bridge of Long Life, toward the western outskirts of the city. Lee sat in the passenger seat and gave directions as best he could. He wasn't absolutely sure where his cousins lived, but his memory guided him to the right place—a small tile-roofed house in the suburbs, not far from where the neighborhoods gave way to open rice fields. When the jeep pulled to a stop, he hopped out and ran up the stairs. "Sun-ja!" he yelled. "Sun-ja!"

She came to the door, but she was confused and uneasy. Why was a soldier barging into the house? "Sun-ja—it is Bae-suk," he said.

"Your brother." She didn't recognize him in his helmet and military garb. But then a smile creased her face. Though she did not remember much about him, she knew him through photographs and family lore. He was a legend, the big brother who'd escaped to the South. She had grown up wondering if she would ever see him again or if he would remain merely a fading image in a photo album. "Big brother!" she cried, wrapping her arms around his legs.

But he was short with her. "Sun-ja," he said, pointing to the street, where the jeep sat idling. "We must go right now."

"Go where?" She did not understand what was happening in the city, or why her brother had appeared out of nowhere, like a phantom.

"There's no time to explain. We must go to the house. To Mama and Papa."

Sun-ja accepted this, but she wondered about her cousins. Could they come, too?

"No," Lee said. "You must say goodbye now."

One of her cousins, Kyeong-ok, was her age—her best friend and playmate. The two girls were inseparable. Kyeong-ok begged to tag along. Wherever Sun-ja was going, Kyeong-ok wanted to follow. Whether they understood that Sun-ja's family was embarking on a ship, probably never to return, was unclear. But both girls began to sense that this might be a final goodbye.

Kyeong-ok was crying now, wailing, shrieking. Sun-ja was crying, too. Kyeong-ok clutched Sun-ja and wouldn't let go. Finally, Lee had to step in and physically separate the two hysterical girls. He gathered Sun-ja in his arms and hustled to the jeep.

Lee held his crying sister. He had succeeded in accomplishing his errand, but he did not feel successful. All he could think, as the jeep rattled toward the river, was how much he hated this war and the choices it forced people to make.

=

At the entrance to the Bridge of Long Life, the MPs waved them on and they eased through the crowds of refugees waiting on the west side. Lee could see the bitterness and dejection on the refugees' faces, and he felt a claw of guilt at the ease with which he could pass over what was, to his fellow countrymen, an impermeable barrier. The jeep crossed the river and paused at the checkpoint on the east side. Again, the MPs waved them on through. Lee indicated that he and Sun-ja could walk from here, but his supervisor said with a grin that he wasn't stopping here, that he wanted to see this project through to its proper end. They eased through the crowds and worked their way into the inner city.

Soon they pulled up to the house, and Lee and Sun-ja climbed out. Their parents were waiting at the door, overcome with emotion. Once the tears subsided, the whole family fell into titters of relieved laughter. They were beaming, brimming with joy. Even Sun-ja.

"I must go," Lee interrupted. They could hear the jeep's engine rumbling outside.

His father gave him a searching look. "We will see you again," he said. "In the South."

The next day, when Lee went back to visit his boyhood compound, it was deserted. Apparently, everyone in his family had left for the Hungnam docks.

In a few more days, American demolitions specialists would strap dynamite to the Bridge of Long Life, and blow it to bits.

45

WE WALK IN THE HAND OF GOD

Hamhung

For two days, they threaded down the mountains, toward the alluvial plains and the sea beyond. Their boots clomped on the frozen road, beside the murmuring trucks. Along the way, at certain switchbacks, the enemy still presented himself—an ambush here, sniper fire there. But the Chinese, it seemed, had largely given up. These Americans were coming out, and nothing could halt them now.

An Associated Press reporter flying over the procession found it strangely beautiful. "Seen from the air," he said, the march "held both magnificence and pathos. There was a Biblical pageantry about it." The men limped on makeshift splints and canes. Their arms dangled in slings. Their bloody clothing was tattered and ripped by shrapnel. Many had draped themselves in cowls of parachute silk. They were gaunt and greasy and chewed up. They had scales on their flesh. They were hairy, soot-smudged wretches, and they stank like herds of wildebeest.

But they were proud. As they marched toward the sea, a peculiar mood settled over the ranks. It was not sadness, nor triumph, nor relief, though all these emotions were present. The overriding feeling might more accurately be described as insolence, a kind of contempt for the whole wide world. They had seen things, they'd been part of something, that they sensed would live forever. They had lost their innocence on a battlefield that had forced them to locate strengths they didn't know they had. It was a cliché often uttered by participants in great battles, but it was true: They formed their own brotherhood now. The Frozen Chosin, they started to call themselves. The Chosin Few. This fraternal spirit welled up in the form of a war chant that rippled down the line of men:

Bless 'em all, bless 'em all,
The Commies, the U.N., and all
Them Red soldiers hit Hagaru-ri
And now know the meaning of the "U.S.M.C."
So, we're saying good-bye to them all
As home through the mountains we crawl
The snow is ass-deep to a man in a jeep
But who's got a jeep?
Bless 'em all!

They had marched into the mountains, and now that they were marching out, they were different men. "They had been there and back," Marine historian Robert Leckie wrote. "They had been through a hell that blazed and froze by turns, and they were singing their splendid disdain for all those pallid paltry souls who did not go."

David Douglas Duncan, the Time-Life photographer, who walked with the Marines on their way out, wrote that the "shuffle of their feet follow[ed] the rising and falling beat of a tragic rhythm." He found a young Marine from Indiana named Robert Henry, who was sitting by the side of the road, spooning at a C ration of half-frozen beans. Henry had vacant eyes, Duncan said, and "the cold had cut into his face" until "even the look of animal survival was gone." Duncan asked the young man, "If you could be granted any wish, what would it be?" Henry thought about the question for a while. Then, prodding his beans again and adopting an idiom reminiscent of the last war, he said, "Give me tomorrow."

=

The vanguard of the Marines passed through a cleft in the hills and suddenly a panoramic vista opened up: the coastal flats below, extending toward the twin cities of Hamhung and Hungnam, and, beyond that, an armada of ships anchored in the glittering port. The Marines wept tears of joy. They had a soft spot for the sea—it was their natural habitat. Down there were hot meals and envelopes from home, hospitals and airfields and tent cities snapping in the salt air. The sight of it seemed to bolster their motivation, and the column moved a little faster.

As they descended, the mercury nudged upward. They began to emerge from a white world of ice into a brown world of mud. By the time they reached Sudong, the Marines found the weather almost balmy. Said one corporal: "When we hit the valley it was like going from Minnesota to Florida. *Boom*, it wasn't cold anymore." This was a most welcome thing—except for those men whose wounds had effectively been sutured by the freezing cold, wounds that now leaked blood again and in some cases showed signs of infection.

Along the Marines' flanks, refugees also jostled down the road. Many thousands of civilians had emerged from the mountains. They were stooped old men, mothers with infants, children whose stony eyes had already seen lifetimes of grief. Some of the refugees had a few belongings strapped to rickety A-frames on their backs, maybe a duck or a chicken, a photograph or some piece of furniture. Some had lost their homes, perhaps their whole villages, to the war. Others were starving and didn't know where else to turn. They followed the Marines out, hoping that a few morsels would come their way—and that if they reached the coast, they might prevail upon the American authorities and hitch a ride out of this war-torn land to settle somewhere in the South.

The refugees, wrote one Marine, "moved like a sullen, brown river, moiling toward the sea." The Americans pitied them but could not let them mix with their own advancing columns. Not only did they slow the procession, but the Marines had already experienced instances of Chinese soldiers, dressed in peasant clothing, melting into crowds of marching refugees, only to hurl a grenade or brandish a hidden weapon.

When the column reached the village of Chinhung-ni, the Marines came into the protective embrace of the Army's Third Infantry Division, elements of which General Almond had sent to meet them. The reservoir country was behind them; their trial was over—everyone could feel it. Now it was a straight shot to Hamhung. Some went by truck, others by train, though, as a matter of pride, many insisted on walking the final fifteen miles.

Many Marines seemed to know that the transports assembled in Hungnam Harbor were intended for them, that the ships had come to evacuate all of X Corps from North Korea, in a sealift that promised to rival Dunkirk in its scale. But perhaps just as many Marines marched toward the coast with the assumption that they would merely be wintering within the safety of Hamhung, that they would establish a stout enclave and resume the offensive in the spring.

For the most ardent of the Marines, this was not only an assumption—it was a distinct desire. After all they'd been through, they couldn't bear the idea of packing up now. They preferred to think of Hamhung as an interlude, not an ending. They still had more fight left in them and remained eager to hear Oliver Smith's next orders. Said one Marine: "I'd follow him to hell because I know he'd get me out."

≡

The possibility that this might be a march to a temporary safe haven, from which Smith's Marines would launch another attack, took some of the sting out of the word *retreat*. Were they, in fact, retreating? Without question, the Chinese had driven them from the field of battle. But most of Smith's men preferred to think of themselves as the victors. "We kicked the shit out of the Chinese the first time we met them, which was at Sudong," boasted Lieutenant Joe Owen, "and we were still kicking the shit out of them when we crossed the treadway bridge. They were surrendering to us, not the other way around. Retreat, you say?"

The way most Marines would come to view it, the decision to push to the Chosin Reservoir had been strategically disastrous. But the battle, once set in motion, had unfolded as an impressive succession of tactical victories. Whether one called it a fighting retreat or an attack in another direction, the First Marine Division had sliced its way through seven Chinese divisions and parts of three others. General Song's Ninth Army Group had been rendered ineffective as a fighting force. Two of his divisions were entirely destroyed, never to be seen on a battlefield again.

The Marines had inflicted astonishing casualties: Song's forces had suffered an estimated 30,000 killed in action and more than 12,500 wounded. The Marines, on the other hand, had lost 750 dead, with 3,000 wounded and just under 200 missing. Though Mao could technically claim a victory in the Chosin engagement—and he loudly did—it was a Pyrrhic one at best. Mao, said one account, had "committed the unforgivable sin, of defeating an enemy army while failing to destroy it."

Back home, the story of the Marine breakout was already widely being described as an "epic." The public thrilled to the news of Smith's smashing his way out of the Red trap. "The running fight of the Marines," said *Time,* "was a battle unparalleled in U.S. military history. It had some aspects of Bataan, some of Anzio, some of Dunkirk, some of Valley Forge, some of the 'Retreat of the 10,000' as described in Xenophon's *Anabasis.*" In a week of catastrophically bad news from Korea, virtually everyone in America seemed to rally around this one stirring headline. In Washington, President Truman extolled the Chosin campaign as "one of the greatest fighting retreats that ever was. Those Marines have old Xenophon beat by a mile." Truman's liaison in Korea, General Frank Lowe, agreed, and went so far as to call the First Marine Division "the most efficient and courageous combat unit I have ever seen or heard of."

General Smith was much more modest. The main reason Chosin had succeeded, he maintained, was that "there was a plan, and that plan was carried out." Smith said, "I knew what we had to do, and I never for a moment doubted we would do it." It was also true, he had to admit, that his division was extraordinary—"I have never commanded a finer body of officers and men," he said. But something else, something ineffable and transcendent, had taken place at Chosin, Smith felt. "Some said our successful breakout was a miracle," he would later reflect. "Some attributed the result to the individual bravery and determination of the officers and men. But more than that was required. One of my regimental commanders summed it up in this fashion: He stated that he was not a religious man, but he felt that we had walked in the hand of God."

Military historians would place much of Chosin's success squarely on the shoulders of one man: Oliver Smith himself. S. L. A. Marshall, a noted Army combat historian, would come to regard Smith as one of American history's most underappreciated generals. "The Chosin Reservoir campaign is perhaps the most brilliant divisional feat of arms in the national history," Marshall wrote. "Smith made it so, through his dauntless calm. In battle, this great Marine had more the manner of a college professor than a plunging fighter. But our services have known few leaders who could look so deeply into the human heart. His greatest campaign is a classic which will inspire more nearly perfect leadership by all who read and understand that out of great faith can come a miracle."

≡

General Oliver Smith came down the mountain on December 10. At a temporary headquarters in Hungnam, he was forwarded a sheaf of letters from his wife, Esther—letters that first conveyed worry, then genuine alarm, then overwhelming relief and elation. "To say that I am proud of you is putting it mildly," she wrote in a later note. "Your march is being called many things—'an attack in reverse'—'a fighting march to the sea,' etc. But the description I like best is that it is a 'splendid moral victory.' I think it was just that and I am very grateful." In another letter, Esther wrote, "My admiration for your calmness and self-containment grows over the years."

Smith had lost a fair bit of weight, and several weeks of sleep deprivation wore on his face. But upon reaching the coast, he looked in fine fettle. The adversities of war seemed to agree with him. "I never felt better in my life than during the period I spent in the mountains," he later insisted. "Oftentimes when you are busy enough you forget about your health."

Conspicuously, Douglas MacArthur was not on hand to greet Smith or his battle-worn Marines when they reached the sea. It seemed not to register with the supreme commander that the nightmare the Marines had just passed through bore any relationship to him, or decisions he had made. The best MacArthur could muster was a perfunctory photo op, on December 11, at Yonpo Airfield, where he pronounced the evacuation a success, then flew straight back to Japan. By now, Smith had learned that MacArthur had decided to pull everything out of Hamhung, to regroup in the South at a place near Pusan, and to fight another day. This was not Smith's wish—he believed, like Almond, that Hamhung-Hungnam could be held indefinitely, or at least until the spring. It was odd for Smith to find himself, for once, in full agreement with the X Corps commander. In fact, during the past week, he and Almond had had, if not a rapprochement, then at least a cooling of tempers. Almond, to his credit, had impressively shifted gears, helping to execute the withdrawal with the same eager resourcefulness he had invested in his ill-advised race for the Yalu. At the least, his energy and enthusiasm were infectious. "Although Almond and I did not always see eye to eye," Smith wrote, "I will say this for him: when we had come out of the Chosin Reservoir, he was still full of fight."

Almond had nothing but praise for the Marines and the way they had acquitted themselves at Chosin. With surprising effusiveness, Almond said this of Smith's division: "No more gallant men ever fought the battles of our country in any of its wars."

But despite Smith's and Almond's preferences, the fact was, X Corps in its entirety was pulling out of North Korea—and the First Marine Division was leaving almost immediately. Soon after the Marines reached Hungnam, they were told to prepare to board ships. Almond had decided that, given the tribulations they had suffered at the reservoir, the Marines should evacuate first. They had earned that right.

Over the past few days, the beachhead at Hungnam Harbor had become an immense depot of military supplies—Inchon in reverse. Along the piers, the mountains of ammunition and fuel, vehicles and medicines, grew by the hour. Waiting to accept the many tons of freight, as well as the ninety thousand troops of Almond's X Corps, a hundred vessels sat at anchor in the harbor approaches: Navy ships, merchant ships, tank landing ships, transports, and sundry other craft. Assembling this diverse rescue fleet had been a herculean logistical effort—the result of desperate scrounging, at enormous expense, under brutal deadlines.

Around Hamhung-Hungnam, units of the Third and Seventh Infantry Divisions had established a twenty-mile security perimeter, which was designed to contract as the evacuation proceeded; when the last ships departed the harbor, later in the month, the tightening perimeter would vanish altogether. The Chinese made immediate attempts to probe and harass the lines—in one instance, disguised in American uniforms taken from the dead. But with batteries of Army howitzers pounding the hills to the west and the north, the Chinese generally kept their distance while the sealift continued.

≡

By December 12, all three of Smith's regiments had streamed down to the docks. First came Litzenberg's Seventh, then Murray's Fifth, then Puller's First. By the next day, the entire division—22,215 Marines in all—were safely aboard their vessels and settling in for the voyage. On some of the ships, they peeled away their filthy, encrusted clothes and awarded themselves the extravagant luxury of a hot shower. They were told they could use all the water they wanted. They stood under the spigots until their skin turned red, until they'd drawn the chill from their bones. Some men would recall that this was the first time they had been able to weep without shame, their tears mixing discreetly with the water and swirling down the drains.

One of the Marines, Private First Class Jack Wright, recalled, "When I dried off, someone brought me a hot cup of coffee. I dressed in my new clothes, and they asked me what I wanted done with my old ones. I said, 'Like all dead things, *bury them.*'"

Navy doctors boarded the ships and inspected the men. Hunting for signs of gangrene, they separated out the worst cases. "They cut our boots off," remembered Sergeant Sherman Richter. "A doctor walked down the line looking at frostbitten toes, saying: 'Treatment. Treatment. Amputate. Treatment. Amputate. Treatment. Treatment.'"

Before reporting to his own ship, General Smith led a memorial service at the Hungnam cemetery, where scores of Marines lay buried in newly dug graves. He wore a heavy winter parka that disguised how much weight he'd lost during the battle. After the chaplains had made their remarks—"May hate cease and wars be forever ended," said one—Smith stepped forward to speak. He had removed his helmet, and his white hair was luminous against the fresh mounds of red clay.

"On an occasion like this," Smith said, "words are inadequate to express our feelings. These men who lie here fought and died far from home in support of a principle. The memory of what they did here will remain with us always." Smith tarried by the long rows of crosses while a bugler played taps, drawing out the last plangent notes until they were overtaken by the sound of the surf.

Then Smith boarded the USS *Bayfield,* an attack transport, and prepared to set sail for the south. On the far snow-sugared ridgelines, the Chinese armies were massing. The *Bayfield* weighed anchor on December 15 and slipped out of the harbor to join the convoy of other Marine transports. Soon the ships were cruising through the Sea of Japan, and the hills of North Korea slipped away. In his quarters, Smith began to compose a Christmas message to his men. "We do not know what the future holds," he wrote, "but we know that we can face it with the confidence Marines always have in the future. We have much to be thankful for. We have emerged from a supreme test with our spirit unbroken."

IN THE PANTHEON

After the ninety thousand men of X Corps were loaded onto ships and evacuated, the American commanders made a momentous decision: They would turn their attention to rescuing as many North Korean civilians as could possibly be crammed into the remaining vessels in the harbor. The vast humanitarian effort was approved by General Edward Almond, whose officers directed nearly one hundred thousand refugees to be squeezed into the ships—the merchant marine freighter SS *Meredith Victory* alone held more than fourteen thousand. These civilians would be carried to a new life in the South.

Then, in the last week of December, demolitions experts set explosives all along the port. Naval guns were also trained on the city. In a gargantuan display of pyrotechnics, the harbor was destroyed, to prevent anything of use from falling into the hands of Chinese and North Korean troops, as they descended upon the city.

The last Americans left Hungnam on Christmas Eve. "They have gone," said General Song. "We could not stop them."

=

For his valor at Fox Hill, Private **Hector Cafferata Jr.** was awarded the Congressional Medal of Honor from President Harry Truman. (His close buddy Private **Kenneth Benson** received the Silver Star.) When Cafferata learned he'd won the Medal of Honor, he asked if they could just send it to him in the mail—but the Marine Corps prevailed upon him to attend the White House ceremony. "I'm not a hero," he later insisted. "I hate heroes. And I hate medals. There's plenty guys who did more than me. They didn't get recognized. All they got was dead." Having recovered from his battle injuries, Cafferata returned home to New Jersey, where he sold hunting equipment, worked for the state fish and wildlife division, and operated a tavern. He died in Venice, Florida, in 2016, at the age of eighty-six.

As part of the historic Hungnam Evacuation, **Lee Bae-suk**'s entire family was safely settled in South Korea. Lee served with United Nations forces until the war's end, when he resumed his medical studies in Seoul. He became a radiologist and, after emigrating to the United States, enjoyed a prominent medical career in Ohio. Dr. Lee and his wife, Mi-yong, who was rescued from Hungnam with her own family aboard the SS *Meredith Victory*, live in Cincinnati. Today, more than a million South Koreans trace their lineage back to survivors rescued during the Hungnam Evacuation.

Ensign **Jesse Brown**'s body and plane were never recovered. A few days after the crash, a Navy pilot struck it with a canister of napalm and the wreck went up in flames—an aviator's funeral pyre. In recognition of his attempt to save Brown, **Thomas Hudner** was awarded the Congressional Medal of Honor. As President Truman presented the award in Washington, Daisy Brown, Jesse's widow, looked on in tears. Hudner, who served in the Navy until 1973, traveled to North Korea and negotiated unsuccessfully with the government in Pyongyang to have Brown's remains located and returned. Hudner died in Concord, Massachusetts, in 2017, at the age of ninety-three, and is buried in Arlington National Cemetery. A Navy destroyer, the USS *Thomas Hudner*, is named in his honor.

In recognition of his leadership on Fox Hill, Captain **William Barber** was also awarded the Congressional Medal of Honor. After healing his shattered hip at a military hospital in Japan, he returned to a career in the Marines, serving in Vietnam and retiring in 1970. He died in 2002 in Irvine, California, aged eighty-two.

For his exploits in the battles of Sudong and Chosin, **Chew-Een Lee** won the Navy Cross, the Silver Star, and two Purple Hearts. Subsequent efforts by comrades, friends, and family to award him the Medal of Honor have so far been unsuccessful. Lee retired from the Marines as a major in 1968. He died in 2014 in Washington, D.C., at the age of eighty-eight.

Following a long stint in military hospitals, recovering from serious facial wounds and undergoing elaborate dental reconstruction, Lieutenant **John Yancey,** having won the Navy Cross and a Silver Star, returned to his home state of Arkansas to run his liquor store and raise a family. He later made an unsuccessful bid for the state senate, opposing segregationist Governor Orval Faubus's machine. During the Vietnam War, Yancey tried to reenlist in the Marines, but he failed his medical test; the examiners noted that he did not have the number of teeth required to play a combat role. He wrote in reply, "I wasn't planning on *biting* the sons of bitches." Yancey spent much of his later life in Mexico and died at sixty-eight, in 1986.

Private First Class **Jack Chapman** spent thirty-three months as a prisoner of war in North Korea—all of them with the bullet that had nearly killed him still lodged in his skull. (He finally had it removed in 1960; he still keeps it as a relic.) Chapman was released in 1953 and returned to the United States, pursuing a career in the Air Force, then serving as the chief of police at a college campus. Retired, he lives with his wife in Santa Fe, New Mexico.

After being airlifted from his ordeal on the east side of Chosin, Army private **Ed Reeves** spent years in and out of military hospitals. Due to the severe frostbite he suffered, both of his feet and all of his fingers were amputated. He married shortly after his Korean service, had six children, became a computer programmer, and spent many years as a missionary in South America. He died in 2010 at age seventy-eight, in Prescott, Arizona.

In the months after China's intervention in the Korean War, General of the Army **Douglas MacArthur** made increasingly strident calls for dropping atomic bombs on Beijing and other Chinese cities and even suggested sowing a permanent radioactive zone, a kind of nuclear fence, along the Manchurian border. In April of 1951, he was relieved of his command by President Truman. "I didn't fire him because he was a dumb son of a bitch, although he was," Truman later said. "I fired him because he wouldn't respect the authority of the president." MacArthur died in 1964. Truman served out his term and died in 1972 in Kansas City. He is buried on the grounds of the Truman Presidential Library and Museum, in Independence, Missouri.

General **Edward Almond** continued to lead the X Corps in Korea, participating in the resounding defeat of several Chinese offensives. He left Korea in 1951 to serve as president of the U.S. Army War College. After retiring from the Army two years later, he became an insurance executive in Alabama. He died in 1979, aged eighty-six, and is buried in Arlington National Cemetery.

The **Korean War** carried on until July 1953, when an armistice was signed. The conflict had ended in a virtual stalemate: The boundary between North and South Korea stood essentially where it had when the hostilities began. A "demilitarized zone" was established not far from the thirty-eighth parallel. According to the Pentagon, 33,651 Americans had died fighting in the war, as did 180,000 Chinese. An estimated 2.5 million Korean civilians lost their lives. Technically, the war is still not over. The armistice provided for "no final peaceful settlement." The two Korean nations have been poised on the brink of war ever since.

Presiding over the First Marine Division in Korea until May of 1951, **Oliver Prince Smith** was appointed commander of the Marine base at Camp Pendleton, California. He retired in 1955 with the rank of four-star general and died in his sleep on Christmas Day 1977, at his home in Los Altos, California, surrounded by his rose gardens and the fruit trees he tended each day. A widower of thirteen years, Smith was eighty-four.

A Chosin survivor, Captain William Hopkins, had this to say about Smith: "I am forever grateful that Oliver P. Smith commanded . . . at Chosin. He embodied all of the features required by Sun-tzu: 'Wisdom, sincerity, humanity, courage, and strictness.'" Said Benis Frank, chief historian of the Marine Corps: "I am certain that somewhere there is a pantheon for Marine heroes where General Oliver Prince Smith holds an honored place."

Smith's granddaughter, Gail Shisler, observed that at the time of his death, behind an old green leather chair in his book-lined study, was a hand-painted map of the Chosin Reservoir, "showing the route of his division's epic battle to the sea." She also noted this detail: "His gardening boots, the same boots that he had worn coming out of the reservoir, stood under a bench by the back door."

ACKNOWLEDGMENTS

Certainly the most profound and rewarding part of researching this book was spending time with scores of veterans of the Chosin battle. "The Chosin Few" are an exceedingly stoic and often irascible group of men who know the deepest meanings of suffering. I met many in their homes and many at reunions—traveling, in the end, to more than twenty states. Most of these men had served under General Smith in the First Marine Division, although many were Army, Navy, or Air Force veterans, and a few were civilians who had gotten caught in the fray. Their stories are the lifeblood of this book. I can't thank them enough—for their good spirit, for their candor, and for their willingness to revisit memories that, in many cases, they would rather leave undisturbed. I simply cannot imagine what they endured all those years ago in that icy crucible in the mountains of North Korea.

Among the veterans I interviewed: Frank Abasciano, Harrison Ager, Robert Arias, Robert Ayala, Thomas Beard, Lacy Bethea, Woody Birckhead, Rayburn Blair, Richard Bonelli, Fred Borowiec, Harry Burke, Hector Cafferata, Franklin "Jack" Chapman, Linus Chism, Watson Crumbie, Norman Deptula, Ken Dower, Larry Elwell, Robert Ezell, Patrick Finn, Sam Folsom, Robert Gaines, Raymond Garland, John Edward Gray, Max Guernesey, Bill Hall, Robert Harbula, Robert Harley, Robert Himmerich y Valencia, Tom Hudner, Dr. Ralph Jacobs, Bob Johnson, Eugene Johnson, Manert Kennedy, Richard Knoebel, John Y. Lee, James McInerney, Charles McKellar, Ralph Milton, Joseph Mordente, Bill Paganini, Kenneth Popp, General Edward Rowny, Elmer Schearf, Elliott Sortillo, Fred Sozio, William Steele, Dr. James Stewart, Duane Trowbridge, Warren Wiedhahn, and Dr. Stanley Wolf.

I'm especially thankful to Dr. Lee Bae-suk and his wonderful wife, Mi-yong, who generously hosted me in their home in Cincinnati, treated me to an outrageously good Korean meal, and shared the extraordinary story of their war years and their flight from North Korea. I had originally met them at a Chosin reunion in Springfield, Missouri. One could not invent an algorithm that captured all the eleventh-hour twists and lucky intersections of their eventful, beautiful, and sometimes tragic family history. Let it serve as a powerful reminder to any who may have forgotten that the story of the Korean War is only secondarily an American one.

I'm indebted to historian Dr. Xiaobing Li, at the University of Central Oklahoma, a native of Beijing, for his many insights into the Chinese perspective on the battle. Dr. Li kindly shared an early draft of his manuscript, *They Came in Waves: The Chinese Attack on Chosin,* the first book that's been written on the Chinese experience of the battle. In Taiwan, I must thank my stalwart researcher Simon Lin, who helped me locate a number of Chinese veterans of the battle at the reservoir. Also helpful was Hong Kong–based historian David Chang, an authority on Chinese prisoners of war during the Korean War. My good friend Professor Y. David Chung, at the University of Michigan, in Ann Arbor, provided many insights into Korean culture and history.

I thank my friend Bill Broyles, a world-class screenwriter and Vietnam Marine, for convincing me to write about Chosin in the first place. I must also thank Bill Miller and the good folks at the Santa Fe Institute for an extremely generous fellowship in 2015, during which many of the ideas and research strategies for this book were conceived. Similarly, my huge thanks to the Ucross Foundation, in Wyoming, for a remote ranch-country writing retreat that proved inspiring and rejuvenating.

In the early going, I was fortunate enough to have struck up a fruitful relationship with documentary filmmaker Randall MacLowry and associate producer Rebecca Taylor at the Film Posse, in Boston, and to have served as a historical consultant for MacLowry's searing war documentary "The Battle of Chosin," which first aired on the PBS program *American Experience* in the fall of 2016. *American Experience* graciously made available the transcripts of all interviews conducted in connection with the making of the film—a substantial oral history archive numbering in the thousands of pages. At WGBH in Boston, I thank Mark Samels, Susan Bellows, John Bredar, and Jim Dunford.

I want to thank the legendary *Life* magazine photographer David Douglas Duncan, whose images of the Chosin battle are iconic and unforgettable. Duncan, then one hundred years young (he has since passed away), hosted me in his home in Provence, France, and shared his memories of the battle. My thanks to former Time-Life editor Richard Stolley for connecting me with Mr. Duncan. At the Union Jack Club, in London, I was fortunate enough to attend a reunion of the last surviving members of 41 Independent Commando of the British Royal Marines. In connection with that festive event, I must especially thank Chosin veterans Gordon Payne, Kenneth Williams, John Underwood, and Cyril Blackman, as well as Lin Blackman for making it all possible.

In Baton Rouge, Barbara Broyles, daughter of the late Colonel Don Faith, kindly made available her dad's personal papers, photographs, and relics. I'm indebted to Jamie Polk, in Santa Fe, who shared the revealing unpublished war letters of his late father, Lieutenant Colonel James Polk, a high-ranking Army intelligence officer within MacArthur's headquarters in Tokyo. I must also thank the family of the late John Yancey— especially Stuart Yancey, Anne Yancey, and Laura Neve—for sharing copious stacks of family memorabilia and correspondence and for showing me a good time in Little Rock. Stephan McAteer, at the MacArthur Museum of Arkansas Military History, was generous with his time in leading me through the Yancey Collection.

I was blessed to have several dogged researchers during the early stages of this project. I especially thank Gillian Brassil and Joel LeCuyer for their smart and indefatigable efforts. Also, Allison Goodwin did invaluable research into the Chinese perspective of the battle and helped me locate some of the last living Chosin veterans in Taiwan and mainland China. Graham Sides provided great help in organizing my source material and assembling my bibliography.

I'm grateful to Julie Precious (producer and director of the excellent Chosin documentary *Task Force Faith*), who generously shared her contacts and insights on the battle. Thanks also to Colonel Ashton Ormes, an expert on Chosin. Charmaine Francois-Griffith was extremely helpful in connecting me with Army veterans of the battle. Former congressman Pete McCloskey of California was kind enough to share his memories of Korea and to connect me with several Marine survivors of Chosin. Adam Makos, author of *Devotion*, introduced me to several intrepid Korean War aviators and graciously read and commented on portions of my manuscript.

At the Truman Presidential Library, in Independence, Missouri, I particularly want to thank archivists Sam Rushay and Randy Sowell. Jim Zobel, at the MacArthur Memorial Library and Archives, in Norfolk, Virginia, was tremendously helpful. My research at the Marine Corps Archives, in Quantico, Virginia, was terrifically productive, thanks to the good efforts of Annette Amerman, Dr. Fred Allison, and Dr. Jim Ginther. Also at Quantico, I had the privilege of meeting the wonderful Gail Shisler, Oliver Smith's granddaughter and the author of the definitive biography of the general. I thank her for her intelligent insights and good cheer—as well as her willingness to read over an early version of the manuscript. What a fine man she had to call "Grandpa."

Lots of generous people hosted me during my far-flung research travels and writing retreats. In this regard, I especially want to thank George and Cindy Getschow, in Texas; the de Bontin family, in Taos; Jessica Goldstein and Peter Breslow, in Washington; Alyssa Brandt and Jay Stowe, in Cincinnati; and Sarah and Ben Fortna, in London. Bill Banowsky supported this project in numerous ways. Nick Ragland kindly read an early draft of my manuscript. The eagle-eyed Will Palmer improved this book in ways large and small. My thanks, also, to Giovanni Orlando for his tireless efforts while I was immersed in this project. My gratitude to Gary Oakley for his keen photographic eye and to Dan Dearborn for the formidable desk. Thanks also to Jamie Wiedhahn of Military Historical Tours for a great veterans trip to South Korea.

My association with Colorado College, where I served as journalist in residence and visiting professor while researching and writing much of this book, proved a substantial inspiration. I especially thank Dr. Steven Hayward and President Jill Tiefenthaler for their friendship and support, as well as Ian and Susan Griffis and Anne and Dave Hanson.

My longtime friend Sloan Harris, at ICM, is my rock and tireless advocate. I can't thank him enough for his support during this project—and through all my books. Also at ICM, my heartfelt thanks to Heather

Karpas, Heather Bushong, Alexa Brahme, and Will Watkins. My editor at Doubleday, Bill Thomas, is quite simply a force of nature. This book would not have happened without his unstoppable energy and passion. Also at Doubleday, I must thank Margo Shickmanter and the effusively exclamatory Todd Doughty, who believed in this project from the start, and Lydia Buechler and John Fontana.

My wife, Anne Goodwin Sides, is a saint for encouraging my wanderings and scribblings. Her excellent taste, sound judgment, and good cheer always keep me on solid ground.

A NOTE ON SOURCES

This narrative springs largely from documents found in the Marine and Army archives, from various repositories of oral history, and from my own interviews with dozens of Chosin veterans. But the existing literature of the Chosin Reservoir battle is vast, vivid, and varied, and I was fortunate enough to build on the solid and expansive foundation provided by many excellent published works. Chosin is probably the most richly documented battle in the Korean conflict, if not in all of post–World War II American arms. And for good reason: It is one of the most highly decorated clashes in our country's history. Few battles can boast so many notable instances of individual courage, carried out on such inhospitable terrain, in such impossible weather, in conditions of such intimate combat, against such overwhelming numerical odds. The extremity of the predicament brought to the fore a naked survival instinct, a ferocious camaraderie, and a rare improvisational spirit.

It would be a lengthy and impractical exercise for me to cite all the fine books that informed my narrative, but I'd like to single out a few. Without question, Martin Russ's *Breakout* remains the definitive account of the battle for anyone interested in a deeper level of detail on the engagement as a whole, on individual unit movements, and on the day-to-day experience in the field. A decorated Marine, Russ was a sharp observer, a pungent writer, and an exhaustive researcher, and his account holds up exceptionally well after twenty years. Robert Leckie's *March to Glory* may seem a bit purple to modern sensibilities, but there is no better way to get an immediate and visceral feel for the battle than to read this vigorous and spirited little volume. Eric Hammel's *Chosin: Heroic Ordeal of the Korean War* is a rewarding document, chock-full of interesting original detail and first-class reporting. Historian and Korean War veteran Stanley Weintraub has dealt with aspects of the battle in two excellent books, *A Christmas Far from Home* and *MacArthur's War*. For the Chinese perspective, I found three books especially helpful: *Enter the Dragon*, by Russell Spurr; *Uncertain Partners*, by Sergei N. Goncharov, John W. Lewis, and Xue Litai; and *They Came in Waves*, a groundbreaking but (as of this writing) unpublished manuscript by Chinese American historian Xiaobing Li.

Those interested in understanding the larger Cold War geopolitics, as well as the colliding egos and ideologies that set the stage for Chosin and related battles, must rely on David Halberstam's quintessential *The Coldest Winter*, a sweeping and minutely detailed narrative by the late, great journalist and historian, whom I had the privilege of meeting a few years before his death. For the grunt's-eye view, probably the best account is Joseph Owen's *Colder Than Hell*, a raw and honest memoir of a single rifle company's experience on the battlefield. One of the other insightful, if little-known, veteran accounts, I found, was William Hopkins's *One Bugle, No Drums*. Two other indispensable works dealing with the Marine point of view of the overall battle situation are *Frozen Chosin: U.S. Marines at the Changjin Reservoir*, by Edwin Simmons, and *U.S. Marine Operations in Korea, 1950–53*, volume III: *The Chosin Reservoir Campaign*, by Lynn Montross and Nicholas Canzona.

Anyone seeking to understand what happened at Fox Hill should look no further than *The Last Stand of Fox Company*, by Bob Drury and Tom Clavin. This is an excellent and deeply reported narrative of what took place on that charged piece of real estate. The story of Jesse Brown and Tom Hudner is beautifully captured in Adam Makos's *Devotion*, the definitive account of the serendipitous friendship of these two remarkable aviators. Those wanting to know more may wish to consult *The Flight of Jesse Leroy Brown*, by Theodore Taylor, and also *Such Men As These*, by David Sears. The tragic ordeal suffered by the Army units east of the reservoir is exhaustively depicted in Roy Appleman's meticulous *East of Chosin*. For a more personalized account of the Army's travails there, I recommend John Edward Gray's candid autobiography, *Called to Honor*. Survivor Ed Reeves described his brutal experience east of the reservoir in a heartfelt and excruciating narrative, *Beautiful Feet & Real Peace*.

The impeccable character of General Oliver Prince Smith has been explored in several books, but the best source, by far, is Gail Shisler's *For Country and Corps: The Life of General Oliver P. Smith*. Shisler is Smith's granddaughter, but in her meticulous biography she succeeds in treating her subject with discipline and objectivity, letting Smith's humanity show through. An excellent analysis of Smith's battlefield performance can be found in Thomas Ricks's *The Generals: American Military Command from World War II to Today*.

A number of first-rate journalists and photojournalists were on hand to cover the Chosin campaign in person. Two that bear mentioning are *New York Herald Tribune* reporter Marguerite Higgins, who collected her observations in an engaging memoir called *War in Korea: The Report of a Woman Combat Correspondent*, and Time-Life photographer David Douglas Duncan, whose haunting photos and essays were published in an evocative, if somewhat idiosyncratic, volume entitled *This Is War!* One of the very best journalistic accounts of the Korean conflict, and one on which I particularly relied for the story of Inchon-Seoul, is *Cry Korea*, by the gifted British journalist Reginald Thompson.

In terms of what might be called "revisionist" scholarship of the Korean War, probably no one in recent years has done more to place the conflict in the larger context of Korean history and culture than the University of Chicago's Bruce Cumings, and no one has been more frank in describing the brutal and, at times, gratuitous excesses of the American air campaigns unleashed upon North Korea. Cumings's *The Korean War: A History* remains an essential work. Also in the revisionist school is I. F. Stone's *The Hidden History of the Korean War, 1950–51*, a strange, disturbing, and at times brilliant book by the legendary investigative contrarian that questions much of the conventional wisdom about the war.

The Chosin Reservoir battle has also been broadly explored in the pages of fiction. At least five novels dealing with the Chosin campaign bear mentioning: *The Marines of Autumn*, by James Brady; *The Frozen Hours*, by Jeff Shaara; *Retreat, Hell!*, by W. E. B. Griffin; *The Savior*, by Nick Ragland and Tom Schwettman; and *The Coldest Night*, an elegiac work of literary fiction by Robert Olmstead.

NOTES

PROLOGUE: MORNING CALM

3   urine and rotten fish heads: Heinl, *Victory at High Tide*, 104.

3   slime of the mudflats: See Fehrenbach, *This Kind of War*, 165.

3   damaging tanks of butane: Heinl, *Victory at High Tide*, 98.

3   targets towed from the fantails: Davis, *Story of Ray Davis*, 98.

4   a full-scale typhoon: For a good account of the typhoon, see Manchester, *American Caesar*, 578.

4   First came the destroyers: For a full account of the ships involved, see Heinl, *Victory at High Tide*, 89.

5   The bullets whined and smacked: See Higgins, *War in Korea*, 45.

BOOK ONE: SEOUL

1   THE PROFESSOR

10   "load jeeps with their decorations": Leckie, *March to Glory*, 34.

10   "strongest division in the world": Russ, *Breakout*, 6.

10   "a public relations war": Smith to Esther Smith, September 20, 1950, Box 2, General Oliver Smith Collection.

11   "born actor": Smith to Esther Smith, August 23, 1950, Box 3, General Oliver Smith Collection.

11   "pomposity of his pronouncements": La Bree, *Gentle Warrior*, 114.

11   "ascetic thinker and a teacher": Heinl, *Victory at High Tide*, 36.

11   the death of his only sibling: For more on this family tragedy, see Shisler's account in *For Country and Corps*, 74.

11   "a professional killer": Brady, *Marines of Autumn*, 81.

12   "school man and a staff man": Heinl, *Victory at High Tide*, 36.

12   "one of those rare men": Leckie, *March to Glory*, 41.

12   "If you think of a forceful person": Shisler, *For Country and Corps*, 128.

12   "He is a very kindly man": Lowe to Truman, September 15, 1950, President's Secretary's Files, Truman Presidential Library.

14   "I can handle it": Ibid., 64.

14   "the Oriental mind": Stanton, *America's Tenth Legion*, 62.

14   "We must strike hard": Manchester, *American Caesar*, 575.

14   "seal off the entire southern peninsula": Ibid.

15   "I can almost hear": MacArthur, *Reminiscences*, 350.

15   "a 5,000-to-1 gamble": Langley, *Inchon Landing*, 39.

16   "regard for his own divinity": Brady, *Marines of Autumn*, 6.

16   "He pulls no punches": Smith to Esther Smith, September 14, 1950, Box 21, General Oliver Smith Collection.

16   "MacArthur was God": Smith, "Oral Reminiscences of Gen. O. P. Smith," Marine Corps History Division, 9.

16   "more than confidence": James, *Years of MacArthur*, 467.

17   "I shall always be grateful": Edward Almond to Margaret Cook Almond, September 18, 1950, Box 77, General Edward Almond Collection, Army Military History Institute.

17   repeatedly called him "son": La Bree, *Gentle Warrior*, 107.

17   "My first impression": Ibid.

18   "no organized enemy": Ibid.

18   "fantastically unrealistic": Ibid., 111.

18   "formidable physical difficulties": Shisler, *For Country and Corps*, 145.

18   "purely mechanical": Heinl, *Victory at High Tide*, 44.

18   "stuff of legends": Halberstam, *Coldest Winter*, 307.

19   "administrative maelstrom": Shisler, *For Country and Corps*, 129.

19   "puffs of smoke": Thompson, *Cry Korea*, 51.

19   "under a yellow pall": Thompson, Ibid., 61.

19   "The quake and roar": Higgins, *War in Korea*, 142.

20   "the crimson haze": Ibid., 145.

20   "white rabbit out of a hat": Pearlman, *Truman and MacArthur*, 105.

20   "Napoleonic pose": James, *Years of MacArthur*, 478.

20   "the happiest moment": Pearlman, *Truman and MacArthur*, 107.

20   "Our losses are light": Knox, *Korean War*, 251.

20   "shone more brightly": Sloan, *Darkest Summer*, 234.

20   "professionals were doing it": Heinl, *Victory at High Tide*, 120.

2   TRAITOR'S HOUSE

22   Looking after them was their cousin: All descriptions, quotations, and details in this chapter are based entirely on my personal interviews with Dr. Lee Bae-suk at his home in Cincinnati, OH, in September 2016.

24   "awkward bedfellow of 'democracy'": Thompson, *Cry Korea*, 109.

24   "the sun of mankind": Harden, *Great Leader and the Fighter Pilot*, 2.

25   "imperialists and their stooges": Ibid., 4.

3   ACROSS THE HAN

29   "organizational genius of the Americans": Thompson, *Cry Korea*, 60.

29   "a burned-out out husk": Ibid., 82.

31   "General Almond had a habit": Shisler, *For Country and Corps*, 151.

31   "no dearth of advice": Ibid., 157.

31   "a 'hard charger'": Bowser is quoted in Simmons, *Frozen Chosin*, 10.

31   "could evoke the thunders": Fehrenbach, *This Kind of War*, 163.

31   "precipitate a crisis": Shisler, *For Country and Corps*, 156.

31   "the most reckless man": Haig, *Inner Circles*, 44.

32   "Ned was aggressive": Blair, *Forgotten War*, 32.

32   "plenty of enemy fire": Smith to Esther Smith, September 20, 1950, Box 2, General Oliver Smith Collection.

33   "Women carried huge bundles": Owen, *Colder Than Hell*, 63.

33   "hardly a brick out of place": Alpha Bowser, Oral History, 229–30, Marine Corps History Division.

34   "the gallant commander": Shisler, *For Country and Corps*, 152.

34   "meant for gallantry in action": Smith to Esther Smith, September 22, 1950, Box 2, General Oliver Smith Collection.

4 BENEATH THE LIGHTHOUSE

36   Even from the depths: As with chapter 2, all scenes, details, recollections, and descriptions in this chapter are drawn from my personal interviews with Dr. Lee Bae-suk.

5 THE BATTLE OF THE BARRICADES

39   "Slowly and inexorably": Thompson, *Cry Korea*, 94.

40   "town shot to hell": Alexander, *Battle of the Barricades*, 36.

40   "so terrible a liberation": Ibid., 24.

40   "crack of bullets overhead": Duncan, *This Is War!* 84.

40   "deployed his men like ferrets": Thompson, *Cry Korea*, 87.

41   "a dirty, frustrating fight": Alexander, *Battle of the Barricades*, 38.

41   "clots" of enemy corpses: Ibid., 33.

41   "whittled us pretty keen": Ibid., 29.

41   "That was for thank you": Higgins, *War in Korea*, 153.

42   "hysterically babbling words": Duncan, *This Is War!* 84.

42   "withered walnut faces": Thompson, *Cry Korea*, 98.

42   "wasn't in the speed of mind": Almond, Oral History, p. 52, General Edward Almond Collection, Army Military History Institute.

42   "always had excuses": Simmons, *Frozen Chosin*, 2.

42   "exasperatingly deliberate": Alexander, *Battle of the Barricades*, 19.

42   "experienced to be believed": Haig, *Inner Circles*, 42.

42   "Curtness was his hallmark": Ibid., 41.

43   "frosty blue eyes": Ibid., 42.

43   "Being from the South": Atkinson, *Day of Battle*, 383.

43   "characteristics of the Negro": Shisler, *For Country and Corps*, 153.

43   "Negro soldiers learn slowly": Atkinson, *Day of Battle*, 383.

43   "He was not a believer": Haig, *Inner Circles*, 43.

45   "probably more trouble": Shisler, *For Country and Corps*, 146.

45   "coordination of fires": Smith is quoted in La Bree, *Gentle Warrior*, 126.

45   "hit the ceiling": Heinl, *Victory at High Tide*, 212–13.

45   "If you'll give your orders": Alexander, *Battle of the Barricades*, 19.

45   "propensity to relieve subordinates": La Bree, *Gentle Warrior*, 126.

46   "Three months to the day": Alexander, *Battle of the Barricades*, 36.

46   "If the city had been liberated": Ibid.

46   "retreating enemy": Shisler, *For Country and Corps*, 160.

47   "on impulse without serious consideration": Ibid., 161.

47   "I can't pursue anybody": La Bree, *Gentle Warrior*, 130.

47   "I understand your problem": Ibid.

6 THE SAVIOR OF OUR RACE

49   "fearless example": Stanton, *America's Tenth Legion*, 112.

50   "grace of a merciful Providence": Alexander, *Battle of the Barricades*, 46.

50   "discharge of the civil responsibility": Thompson, *Cry Korea*, 110.

51   "We admire you": Manchester, *American Caesar*, 583.

51   "Let the sons of our sons": Passages from Rhee's speech are quoted in Thompson, *Cry Korea*, 110.

52   "must and will be torn down": Haig, *Inner Circles*, 49.

52   "a superhuman effort": Halberstam, *Coldest Winter*, 331.

52   "someone ready to give it a try": Ibid., 332.

52   "the sorcerer of Inchon": Pearlman, *Truman and MacArthur*, 105.

52   "no stopping MacArthur now": Ibid., 104.

52   "We want you to feel unhampered": Manchester, *American Caesar*, 584.

53   "I regard all of North Korea": Ibid.

53   "everything that moved": Harden, *Great Leader and the Fighter Pilot*, 6.

53   "The complete destruction": Manchester, *American Caesar*, 586.

54   "too depressing": Shisler, *For Country and Corps*, 166.

54   fresh bouquet for the table: Ibid.

54   "close to lethal": Owen, *Colder Than Hell*, 103.

54   "As to how long we will stay": Smith to Esther Smith, January 24, 1951, Box 2, General Oliver Smith Collection.

55   "I hope we do not have to operate": Smith to Esther Smith, October 1, 1950, Box 2, General Oliver Smith Collection, Marine Corps History Division.

7 GOD'S RIGHT-HAND MAN

57   "Only God or the Government": Gunther, *Riddle of MacArthur*, 13.

58   "sly political ambush": Whitney, *MacArthur*, 395.

58   "I've a whale of a job": McCullough, *Truman*, 801.

58   "queer accident of democracy": *Time*, March 8, 1943.

58   Underneath the brass: Hersey, *Aspects of the Presidency*, 3.

58   "Mr. President!": Manchester, *American Caesar*, 590.

58   a "picture orgy": Truman, *Memoirs*, 364.

59   "I have been worried": McCullough, *Truman*, 802.

59   "stimulating and interesting": Manchester, *American Caesar*, 590.

59   "nothing but courtesy": MacArthur, *Reminiscences*, 361.

60   "Do you mind if I smoke?": Ibid.

60   "a forlorn hope": James, *Years of MacArthur*, 507.

60   "But they are obstinate": Weintraub, *MacArthur's War*, 189.

60   "What will be the attitude of China?": Ibid., 190.

60   "the greatest slaughter": Manchester, *American Caesar*, 592. See also Truman, *Memoirs*, 366.

61   "the most persuasive fellow": McCullough, *Truman*, 804.

61   "his indomitable will": Manchester, *American Caesar*, 595.

61   "and happy landing": McCullough, *Truman*, 808.

8 THE TIGER WANTS HUMAN BEINGS

62   "Another nation is in a crisis": Peng, *Memoirs of a Chinese Marshal*, 472.

62   "No concession could stop it": Ibid., 473.

62   "increasing the arrogance of the enemy": Ibid., 476.

63   "We all have black hair": Harden, *Great Leader and the Fighter Pilot*, 57.

64   China had already issued a fair warning: See Halberstam, *Coldest Winter*, 334–37.

65   imbibe the spirit: For more details on Mao's obsession with river swims, see Li, *Private Life of Chairman Mao*, 157–68.

65   To ward off impotency: Li, *Private Life of Chairman Mao*, 104.

65   "A normal bowel movement": Ibid., 107.

66   "chilled through broken lips": Ibid., 117.

67   "like licking a rock": This detail comes from Li, *They Came in Waves*, 104.

67   "do not use good metal for nails": Coggins, *Soldiers and Warriors*, 312.

67   "Growing up, I had no dreams": Author interview with Huang Zhi in Taipei, Taiwan, January 2018.

67   "I didn't have any idea about Americans": Ibid.

68   "The imperialist criminals were at our door": Author interview with Yang Wang-Fu in Taipei, Taiwan, December 2017.

68   "China, though weak": Li, *They Came in Waves*, 94.

68   "Weapons are an important factor": Ibid., 11.

68   "I will find the enemy's weakness": Ibid., 41.

BOOK TWO: TO THE MOUNTAINS

9   MANY, MANY

72   "It took a curious sort of mind": Brady, *Marines of Autumn*, 62.

72   "ordeal of misery and sickness": Owen, *Colder Than Hell*, 104.

72   "Never did time": Russ, *Breakout*, 11.

74   "literally blown to pieces": Ibid., 15.

74   "a surrealistic scene": Haig, *Inner Circles*, 54.

74   "dissipated to a point of ineffectiveness": Intelligence report quoted in Brady, *Marines of Autumn*, 70.

75   the so-called dragon's back: Salmon, *Scorched Earth, Black Snow*, 320.

77   "I don't have much confidence": Simmons, *Frozen Chosin*, 22.

77   "I got troops scattered": Shisler, *For Country and Corps*, 172.

78   "I questioned his judgment": Bowser is quoted in Simmons, *Frozen Chosin*, 10.

78   "was overly cautious of executing": Almond is quoted in Simmons, *Frozen Chosin*, 35.

78   "He was the cock of the roost": Author interview with Sam Folsom in Santa Monica, CA, on January 24, 2016.

79   "many, many": Russ, *Breakout*, 24.

79   "This information has not been confirmed": Stanton, *America's Tenth Legion*, 161.

79   "There is no positive evidence": McGovern, *To the Yalu*, 54.

10   KING'S ENVOY TO HAMHUNG

80   a nineteen-year-old medical student: Details of this chapter, unless otherwise noted, are drawn primarily from my own extensive interviews with Dr. Lee Bae-suk.

82   "king's envoy to Hamhung": See Choe and Torchia, *How Koreans Talk*, 94.

83   "entrepreneurial king of the peninsula": See Harden, *Great Leader and the Fighter Pilot*, 21.

84   "the black umbrella": For an insightful account of Korean life under Japanese colonial rule, see Kang, *Under the Black Umbrella*.

85   Thousands of young Korean women: A first-rate narrative about Japan's wartime program of enforced prostitution is Hicks, *Comfort Women*.

85   a sadistic medical experimentation program: For more on Unit 731, see Barenblatt, *Plague Upon Humanity*.

86   women began disguising themselves: Harden, *Great Leader and the Fighter Pilot*, 27.

86   a decadent bourgeois sport: Ibid., 5.

86   Then followed the parades: An excellent account of the early days of Communist rule in North Korea is Cha, *Impossible State*.

11   HEROIC REMEDIES
88   120 steps a minute: Hersey, *Aspects of the Presidency*, 37.

88   get "your circulation up": Ibid., 48.

89   It was "clearly established": Walter B. Smith, CIA director, "Memorandum for the President: Chinese Communist Intervention in Korea," November 1, 1950, President's Secretary's Files, Truman Presidential Library.
        90   "committing themselves to full-scale intervention": Ibid.

91   "staying up there from force of habit": Klara, *Hidden White House*, 51.

91   "Heroic remedies": Ibid., 68.

91   clutching a German Luger: My account of the assassination attempt is drawn primarily from Hunter and Bainbridge's excellent work on the subject, *American Gunfight*. See also Donovan, *The Assassins*.

93   "An attempt has been made to assassinate": Acheson, *Present at the Creation*, 459.

93   "a bad scare": Ibid.

93   "A president has to expect": McCullough, *Truman*, 812.

94   "I could organize a better program": Truman to Acheson, November 2, 1950, President's Secretary's Files, Truman Presidential Library.

12   WILL O' THE WISP
96   "like a phony war": Leckie, *March to Glory*, 22.

96   "great energy in the ranks": Owen, *Colder Than Hell*, 113.

96   "stories of Japanese ferocity": Ibid.

96   "bit of a bully": Russ, *Breakout*, 66.

96   "built all of rectangles": Hammel, *Chosin*, 23.

97   "important that we win:" Simmons, *Frozen Chosin*, 13.

97   "The results will reverberate": Russ, *Breakout*, 25.

97   "like flocks of blackbirds": Ibid., 30.

97   "quite a fight": Ibid., 47.

99   "a tiny fellow who smiled": Ibid., 58.

99   "we are fighting a sizeable unit": Smith to Esther Smith, November 6, 1950, Box 2, General Oliver Smith Collection.

99   "ought to make noise": Brady, *Marines of Autumn*, 124.

99   "like a will o' the wisp": Smith, Aide-Mémoire, 745, Box 34, General Oliver Smith Collection, Marine Corps History Division.

100   merely a "bother": Almond to Margaret Almond, November 18, 1950, General Edward Almond Collection, Army Military History Institute.

100   "my pet fascist": Gordon, *Modern History of Japan*, 237.

100   "Anything MacArthur wanted": Halberstam, *Coldest Winter*, 378.

100   "He has no wife": Lt. Col. James Polk to his wife, Josephine Polk, October 12, 1950.

101   "We had the dope": Lt. Col. James Polk, personal letter to his wife, Josephine Polk, December 6, 1950.

102   "too obdurate for them": Haig, *Inner Circles*, 59.

102   "every trick in the book": Russ, *Breakout*, 52.

102   "joked and laughed as we marched": Owen, *Colder Than Hell*, 167.

103   "Shadow and shade": Brady, *Marines of Autumn*, 98.

103   "Beyond each hill": Russ, *Breakout*, 27.

103   "Even Genghis Khan": Ibid., 64.

103   "bandit country": Brady, *The Marines of Autumn*, 116.

103   "mysterious Oriental kingdom": Russ, *Breakout*, 50.

13   BROKEN ARROWS

104   "The son of a bitch": Rovere, *Senator Joe McCarthy*, 12.

105   "Korean death trap": McCarthy is quoted in Donovan, *Tumultuous Years*, 295.

106   A U.S. Navy fighter pilot: See Sears, *Such Men As These*, 95.

107   an even weightier event: For a good account of the Rivière-du-Loup B-50 incident, see Mowat, *Eastern Passage*, and also Septer, *Lost Nuke*.

108   "broken arrow": See Oskins and Maggelet, *Broken Arrow*.

14   A POWERFUL INSTRUMENT

109   the first recruits of the Continental Marines: See Heinl, *Soldiers of the Sea*, 4.

110   "your present gallant and heroic exploits": La Bree, *Gentle Warrior*, 142.

110   "a ferocious little confraternity": Brady, *Marines of Autumn*, 128.

111   "It was not because they were braver": Russ, *Breakout*, 5.

113   "a propaganda machine": Truman to Congressman Gordon McDonough, August 29, 1950, Public Papers of Harry S Truman, 1945–1953, Truman Presidential Library.

113   "a howling beast": Leckie, *March to Glory*, 24.

114   "numb the spirit": Simmons, *Frozen Chosin*, 24.

114   "a disappointing crackle": Brady, *Marines of Autumn*, 118.

114   "The cold was a physical force": Author interview with Frank Borowiec in Chicopee, Massachusetts, July 22, 2016.

114   "no breath to boast in": Leckie, *March to Glory*, 24.

114   "blaspheme the goddamn fools": Owen, *Colder Than Hell*, 188.

114   "If I'd known what the temperature was": Author interview with Harrison Ager in San Diego, CA, on August 20, 2016.

114   "stimulants were required": La Bree, *Gentle Warrior*, 142.

115   When the temperature dropped: On Smith's aversion to extreme cold, see Shisler, *For Country and Corps*, 178.

115   "It seemed with each step": Olmstead, *Coldest Night*, 103.

115   "General Winter": Russ, *Breakout*, 63.

116   "barreling up that road": Ibid., 71.

116   "We're not going anywhere": Ibid.

117   "all of that blood and sacrifice": Shisler, *For Country and Corps,* 88.

117   his letter proved to be profoundly prescient: Portions of Smith's correspondence to Cates have been published in numerous sources, but the letter is quoted in its entirety in La Bree, *The Gentle Warrior,* 145–50.

15   FATTENED FOR THE KILL

119   "like a mining camp": Leckie, *March to Glory,* 82.

120   "a scene of magnificent beauty": Owen, *Colder Than Hell,* 197.

120   "one break we get from the winter": Leckie, *March to Glory,* 43.

121   "What casualties?": See Shisler, *For Country and Corps,* 184.

122   "stirring the air itself": Olmstead, *Coldest Night,* 114.

124   "It was shrewd of Smith": Halberstam, *Coldest Winter,* 435.

124   "as if he didn't care": Ibid.

124   "utterly without fear": Haig, *Inner Circles,* 44.

124   "Their breath vaporized": Ibid., 58.

125   "hit the jackpot": McGovern, *To the Yalu,* 94.

125   "outstanding military achievement": Almond is quoted in Weintraub, *Christmas Far from Home,* 32.

125   killed by a Siberian tiger: Simmons, *Frozen Chosin,* 40.

126   "'vases of this kind'": Haig, *Inner Circles,* 60.

126   "a plush state of affairs": Bowser is quoted in Russ, *Breakout,* 76.

126   "spare as the Marine Corps itself": Ibid., 185.

127   "High in the bitter land": Fehrenbach, *This Kind of War,* 199.

127   "fattened up for the kill": Author interview with Hector Cafferata, Venice, Florida, March 26, 2015.

16   NEVER TOO LATE TO TALK

128   "arrogant imperialist state": For my description of the Chinese delegation's appearance at the United Nations, I am indebted to the excellent account found in Spurr, *Enter the Dragon,* 239–45. Unless otherwise noted, my descriptions here come from Spurr.

129   a "face-to-face struggle": Ambassador Wu is quoted in Spurr, *Enter the Dragon,* 245.

129   "A profound friendship": *New York Times,* November 26, 1950.

129   create a "constant din": Spurr, *Enter the Dragon,* 240.

130   a buffer zone along the Yalu: *New York Times,* November 26, 1950.

130   "never too late to talk": Ibid.

130   "like flying blind": Ibid.

131   "The Great Appalachian Storm": For my description of this historic tempest, I found these two articles especially helpful: Jeff Halverson, "In 1950, an Iconic Storm Blasted Through the Eastern U.S. at Thanksgiving," *Washington Post,* November 22, 2017; and "The Great Appalachian Storm in Historical Context," available on the National Centers for Environmental Information website (www.ncei.noaa.gov).

17   NEVER A MORE DARING FLIGHT

132   "massive compression envelopment": See Manchester, *American Caesar,* 606.

134   the "very audacity": MacArthur, *Reminiscences,* 373.

134 "couldn't lie to the chief": Manchester, *American Caesar,* 606.

134 "I'll stick with the plane": Weintraub, *MacArthur's War,* 233.

134 "death grip of snow and ice": MacArthur, *Reminiscences,* 373.

134 "behind the enemy's lines:" Ibid.

135 "a merciless wasteland': Ibid.

135 "like gods in silver armor": See Weintraub, *Christmas Far from Home,* 12.

135 "phantom which casts no shadow": Marshall is quoted in Manchester, *American Caesar,* 607.

135 "the highest combat effectiveness": Mao is quoted in Li, *They Came in Waves,* 75.

135 "highly competent criminals": Drury and Clavin, *Last Stand of Fox Company,* 263.

135 many of Song's shaggy ponies: Ibid., 66.

136 "as you would snakes": Russ, *Breakout,* 81.

136 "never seen a more daring flight": Willoughby and Chamberlain, *MacArthur: 1941–1951,* 391.

137 "a Greek hero of old": Brady, *Marines of Autumn,* 49.

BOOK THREE: THE RESERVOIR

18 EASY FOR US, TOUGH FOR OTHERS

142 "the plaything of the old men": Olmstead, *Coldest Night,* 108.

142 "Easy for us, tough for others": Geer, *New Breed,* 278.

143 a legendary Marine from Arkansas: My portrait of John Yancey and my description of his exploits on Hill 1282 are drawn largely from the Yancey Collection, on file at the MacArthur Museum of Arkansas Military History. I'm grateful to the members of the Yancey family in Arkansas for generously providing further biographical materials. Also extremely helpful was John Yancey's oral history, available on the Korean War Educator website (www.koreanwar-educator.org). Another helpful source is Yancey's entry in the *Encyclopedia of Arkansas History & Culture,* available at www.encyclopediaofarkansas. Finally, I'm indebted to accounts in four books: Wilson's *Retreat, Hell!,* Hammel's *Chosin,* Leckie's *March to Glory,* and Russ's *Breakout.*

143 "learned his own lessons": Fehrenbach, *This Kind of War,* 244.

143 "before breakfast": *Parris Island Boot,* June 15, 1951.

143 "to decapitate me with a sword": Yancey to Martin Russ, October 2, 1984, Yancey Collection, MacArthur Museum of Arkansas Military History.

144 "we would follow him": Wilson, *Retreat, Hell!,* 63.

144 "the charisma and the steel nerve": Marine Ray Walker, quoted in John Yancey's memoirs, available on Korean War Educator (www.koreanwar-educator.org).

144 "a kind of Valhalla complex": James Claypool to Martin Russ, December 25, 1985, Yancey Collection, MacArthur Museum of Arkansas Military History.

144 "die like Marines!": Russ, *Breakout,* 31.

144 "no chance of being free": Mill, *Principles of Political Economy.*

144 "country fella": Russ, *Breakout,* 111.

145 "rich, crazy Americans": The account of this incident is richly detailed in

the Yancey memoirs on Korean War Educator (www.koreanwar-educator
.org). It is also found in Yancey to Martin Russ, October 2, 1984, Yancey
Collection, MacArthur Museum of Arkansas Military History.

146   the "platoon delinquent": Russ, *Breakout*, 30.
149   "Just do what I tell you": Ibid., 138.

19 BOON COMPANIONS

152   "Their infantry is weak": Portions of the Chinese pamphlet are quoted in
        Drury and Clavin, *Last Stand of Fox Company*, 27.
152   "candy ass": Ibid., 25.
153   "a well-kept grave": Ibid.
153   "hell of a good infantry officer": Ibid.
153   Pancho Villa's banditos: See Douglas Martin, "W. E. Barber, 82, Medal
        Winner in Korea Siege," *New York Times*, April 26, 2002.
154   "Luck in combat is fickle": Quoted in Barber's obituary by Adam Bern-
        stein, "Medal of Honor Winner William Barber Dies," *Washington Post*,
        April 23, 2002.
155   "Jesus Christ": Unless otherwise noted, all battle narrative scenes con-
        cerning Hector Cafferata (including recollected dialogue) on Fox Hill
        derive from my own interviews with him in Venice, Florida, March 26,
        2015. I also relied on a lengthy oral history, videotaped June 29, 2000, on
        file at the Marine Corps Historical Center, in Quantico, Virginia. I found
        The Last Stand of Fox Company, by Drury and Clavin, extremely helpful.
        My passages concerning Cafferata and Fox Hill were further enhanced
        by my personal interviews with three other Fox Hill veterans: Richard
        Bonelli, Bob Ezell, and Harry Burke.
155   "put the bullet where it belongs": Author interview with Cafferata.
156   "The Marine thing": Author interview with Cafferata.

20 EASY COMPANY HOLDS HERE

158   *Son of a bitch, Marines:* Geer, *New Breed*, 279.
158   "a witches' conference": Ibid.
158   "a lunatic's delight": Brady, *Marines of Autumn*, 192.
158   "like a pack of mad dogs": Russ, *Breakout*, 124.
158   the snow had "come to life": Ibid.
159   "There was just so many": Author interview with Robert Arias in San
        Diego, CA, on August 20, 2016.
160   "Lay it on, Ray!": Hammel, *Chosin*, 89.
161   "Stay loose, Marines": Russ, *Breakout*, 142.
162   "Easy holds here!": Fehrenbach, *This Kind of War*, 245.

21 WHERE THE BULLET BELONGS

163   "Somethin's happening": As with chapter 19, unless otherwise noted, my
        account of Hector Cafferata's fight on Fox Hill, including all recollected
        dialogue from the battlefield, is drawn from my own interviews with him.
165   lubricated their weapons with whale oil: See Drury and Clavin, *Last Stand
        of Fox Company*, 90.
166   "chop suey sandbags": Brady, *Marines of Autumn*, 144.
167   "Kill or be killed": Cafferata, quoted in his obituary by Sam Roberts,

"Hector A. Cafferata, 86, Dies; Given Medal of Honor for Korea Heroics," *New York Times,* April 14, 2016.

22   GUNG-HO, YOU COWARDLY BASTARDS

168   Private Stan Robinson lounged comfortably: Robinson's return to the battlefield is captured in Leckie, *March to Glory,* 62.

170   "Gung-ho, you cowardly bastards!": Russ, *Breakout,* 145.

170   cradled his eyeball: Yancey's eye injury is described in Hammel, *Chosin,* 94, and Russ, *Breakout,* 148.

171   "Christ, it's Kennemore!": Leckie, *March to Glory,* 64.

171   A topographical map in his coat: During a filmed interview conducted on December 8, 1950, from his hospital bed in Japan, Yancey displayed the coat and map for the camera. See www.youtube.com/watch?v=4if2 XXnHJQw.

23   WHEN THE LEAD IS FLYING

173   "Was it this bad on Okinawa?": This anecdote is from a videotaped oral history with Cafferata, on file at the Marine Historical Center, Quantico, Virginia.

173   "You look like shit": Author interview with Cafferata.

174   "lower than whale shit": Ibid.

176   "They haven't made the bullet": Barber is quoted in Drury and Clavin, *Last Stand of Fox Company,* 107.

176   "Barber was cool": Author interview with Richard Bonelli in Virginia Beach, VA, on October 15, 2016.

176   one thousand casualties an hour: This statistic comes from Li, *They Came in Waves,* 84.

179   shot him twice in the head: Author interview with Cafferata. This story is also vividly recounted in Leckie, *March to Glory,* 72.

179   "tactical necessities": Drury and Clavin, *Last Stand of Fox Company,* 123.

180   "hugging buddies": This detail comes from Li, *They Came in Waves,* 87.

24   A HOT RECEPTION

182   "These people were here to stay": Bowser is quoted in Russ, *Breakout,* 184.

183   "a hot reception": Ibid., 205.

183   a "pesthouse": Leckie, *March to Glory,* 76.

183   "Maybe it'll inspire us": Ibid.

183   "Apparently they were stunned": Oliver Smith, "Oral History," 222, General Oliver Smith Collection.

183   "Until present situation clarifies": Leckie, *March to Glory,* 76.

183   "I halted the attack": Oliver Smith, "Oral History," 222, General Oliver Smith Collection.

184   "to know when to retreat": See Martin Chilton, "Were it not for Wellington, we'd be speaking French," *Telegraph,* July 13, 2011, https://www .telegraph.co.uk/culture/books.

185   "The inference was": Smith, Korean War Log, 94, Box 34, General Oliver Smith Collection.

185   "quite embarrassed about asking us": Russ, *Breakout,* 229.

186   "pretty much a bastard organization": Shisler, *For Country and Corps,* 198.

186   "left our perimeter dangerously vulnerable": Russ, *Breakout,* 229.

187   "stiffen the collective backbone": Ibid., 195.

187   "That man must be crazy": Shisler, *For Country and Corps,* 198.

188   "That's impossible": Appleman, *East of Chosin,* 106–8.

188   "a bunch of goddamn Chinese laundrymen": McGovern, *To the Yalu,* 125.

188   "What a damned travesty": Russ, *Breakout,* 197.

190   an "exotic concert": Ibid., 263.

190   "a whole field got up on its feet": Ibid., 206.

190   "surf lapping on a beach": Ibid., 208.

190   "chattering of machine guns": Smith, "Looking Back at Chosin."

190   a "witch's clatter": Leckie, *March to Glory,* 92.

25   THE WAR COUNCIL

191   "This headquarters is in a terrible slump": Lt. Col. James Polk to his wife, Josephine Polk, December 9, 1950.

193   "couldn't just passively sit by": Willoughby is quoted in Halberstam, *Coldest Winter,* 477.

193   "not been taken by surprise": Whitney, *MacArthur,* 420.

193   "lost face not just before the entire world": Halberstam, *Coldest Winter,* 475.

194   Almond still held MacArthur's torch: Stanton, *America's Tenth Legion,* 232.

26   AN ENTIRELY NEW WAR

195   "We face an entirely new war": MacArthur's cable is quoted in Donovan, *Tumultuous Years,* 305.

195   "certainty of defeat": Truman, *Memoirs,* 384.

196   "We have a terrific situation": Hersey, *Aspects of the Presidency,* 27.

196   "His mouth drew tight": Ibid., 28.

196   "We have got to meet this thing": Ibid., 30.

197   "an ancient mandarin": George Sokolsky, "These Days," *New York Herald Tribune,* December 3, 1950.

197   "impossible to overestimate the seriousness": Donovan, *Tumultuous Years,* 306.

198   "the Soviet Union as an antagonist": Acheson is quoted in "Notes on NSC Meeting, November 28th, 3:00 PM, the White House," Truman Presidential Library.

198   "without an exit": Manchester, *American Caesar,* 610.

198   "they first make mad": Halberstam, *Coldest Winter,* 476.

198   "a line that we can hold": McCullough, *Truman,* 818.

198   "and get out": Pearlman, *Truman and MacArthur,* 137.

199   "I should have relieved": Truman, *Memoirs,* 384.

199   "The issues that face us": Substantial excerpts of NSC-68, including this quote, can be found at the Digital History website (www.digitalhistory .uh.edu).

199   one of the "grimmest" times: Margaret Truman, *Harry S. Truman,* 492.

199   "World War III is here": Truman's handwritten note, December 9, 1950, President's Secretary's Files, Truman Presidential Library.

BOOK FOUR: RED SNOW

27 YOU WILL ALL BE SLAUGHTERED

203 "Holy Christ": Author interview with Hector Cafferata.

204 "I don't know why I'm here": Author interview.

205 "no possible way we can be relieved": Barber is quoted in Russ, *Breakout*, 223.

205 "heavy attacks again tonight": Ibid.

205 Supply sergeant David Smith: Smith's wounding is described in Drury and Clavin, *Last Stand of Fox Company*, 149, and in Russ, *Breakout*, 222.

206 Captain George Farish: Farish's risky helicopter flight is detailed in Drury and Clavin, *Last Stand of Fox Company*, 150–51.

206 perfect decoys: Benson's ruse is described in Leckie, *March to Glory*, 83.

207 "You are completely surrounded!": Drury and Clavin, *Last Stand of Fox Company*, 157.

207 "Marines, tonight you die!": Ibid.

208 "He was motionless": Leckie, *March to Glory*, 88.

208 "There was a flash": Ibid.

208 and killed them all: Ibid., 87.

208 much more than a bee sting: This detail comes from Martin, "W. E. Barber, 82, Medal Winner in Korea Siege," *New York Times*, April 26, 2002.

209 "Captain Barber, will you surrender?": Drury and Clavin, *Last Stand of Fox Company*, 175.

209 "We're short of warm bodies": Russ, *Breakout*, 251.

209 "You lay down, Moose!": Author interview with Cafferata.

210 "You damn fool!": Leckie, *March to Glory*, 87.

28 KISSING A BUZZ SAW

212 "I am here in the name": The full text of Ambassador Wu's United Nations remarks, from which the quotes in this chapter are drawn, can be found online at "People's China Stands for Peace," on the Internet Archive (https://archive.org/details/peopleschinastan00wuxi).

213 "The real intention of the U.S.": Ibid.

213 "The main burden of my speech": Wu is quoted in Spurr, *Enter the Dragon*, 241.

214 "there was no need for courtesy": Ibid., 245.

214 "kissing a buzz saw": Ibid.

29 MORPHINE DREAMS

215 "Hey Moose, when I die": Leckie, *March to Glory*, 116–17.

216 "a nasty little bastard": Author interview with Cafferata.

216 "would have been a Rhodes scholar": Ibid.

216 "they were haunting me": Ibid.

217 "All right, men": Barber is quoted in Leckie, *March to Glory*, 101.

217 "Men of Fox Company": Messman's loudspeaker announcement on Fox Hill is detailed in Russ, *Breakout*, 267, and in Drury and Clavin, *Last Stand of Fox Company*, 211–12.

218 they shot every one of them: The shooting of the Chinese prisoners was

detailed to me by Cafferata and confirmed by my interviews with several other Fox Company veterans. I have never seen reference to it in print.

218　"Man, I was wild": Author interview with Cafferata.

30　NO SOFT OPTIONS

220　"Where do you put the bayonet?": See the entry for Chesty Puller in Wikiquote (https://en.wikiquote.org/wiki/Chesty_Puller).

220　"What the hell do they think": Leckie, *March to Glory*, 118.

220　"can't get away from us now!": Russ, *Breakout*, 230.

220　"The higher brass": Author interview with Manert Kennedy in Monte Vista, CO, on April 14, 2016.

220　"How many Chinese're": Ibid.

220　a "picturesque" character: Shisler, *For Country and Corps*, 167.

221　"poisonous" country: Salmon, *Scorched Earth, Black Snow*, 321.

222　"a joint-by-joint advance": Leckie, *March to Glory*, 104.

222　"no soft options": S. L. A. Marshall, "CCF in the Attack: Part II," Box 4, S. L. A. Marshall Collection, Army Heritage and Education Center, Carlisle Barracks, PA.

222　"Press on at all costs": Russ, *Breakout*, 234.

222　"we'll give them a show": S. L. A. Marshall, "CCF in the Attack: Part II."

222　"a lot of Chinese bastards": Russ, *Breakout*, 233.

222　"The firing was coming in": Author interview with former Royal Marine Gordon Payne in San Diego, CA, on August 21, 2016.

223　"41 Commando, present for duty": Russ, *Breakout*, 235.

224　Chapman was a scrappy kid: The biographical details and battlefield reminiscences concerning Jack Chapman are drawn almost entirely from my personal interviews with him in Santa Fe, NM, in March 2017. I also relied on Chapman's war memoir, *Cherokee Warrior*, available at Korean War Educator (http://www.koreanwar-educator.org).

225　"Protect us, O Lord": Author interview with Chapman.

227　"we will kill you all": Details concerning McLaughlin's negotiations and surrender come from Hammel, *Chosin*, 205–7; Simmons, *Frozen Chosin*, 68–70; Leckie, *March to Glory*, 111–12, and Russ, *Breakout*, 241–42.

31　ONE-MAN ARMY

229　"eyes that bored into you": Owen, *Colder Than Hell*, 28.

229　"The Chinese think we're roadbound": Leckie, *March to Glory*, 129.

230　"Some fellow Marines": Russ, *Breakout*, 287; Drury and Clavin, *Last Stand of Fox Company*, 242–43.

230　a "bold dash": Davis, *Story of Ray Davis*, 111.

231　"destroy" them: Ibid.

232　"not well developed": Russ, *Breakout*, 13.

232　"hardly a congenial companion": Owen, *Colder Than Hell*, 54.

233　"I tell you, it *works*": This quote is from a videotaped oral history with Lee, recorded on August 24, 2000, on file at the Marine Corps Historical Center.

233　"willingly follow me into hell": Ibid.

233　"I'm not the enemy!": Ibid.

234　"Hey, that's mine!": Ibid.

234 "I wasn't yet convinced": Russ, *Breakout,* 54.

234 "prepared to meet my Maker": Videotaped oral history with Lee.

235 "all the clichés apply": Russ, *Breakout,* 293.

235 "almost a mission impossible": Videotaped oral history with Lee.

32 EVERY WEAPON THAT WE HAVE

236 "Recent developments in Korea": The entire transcript of President Truman's November 30, 1950, press conference, including his opening statement, from which I quote here, is available at the American Presidency Project (www.presidency.ucsb.edu).

33 THE RIDGERUNNERS

239 "just over those ridges": Drury and Clavin, *Last Stand of Fox Company,* 251.

240 "I wanted my men": Russ, *Breakout,* 288.

241 a "long wavering line": Leckie, *March to Glory,* 135.

241 like a mortar shell: Russ, *Breakout,* 295.

241 "Time had no meaning": Owen, *Colder Than Hell,* 234.

242 *"Ching du ma?"*: Drury and Clavin, *Last Stand of Fox Company,* 263.

242 "The voices had an adrenaline effect": Owen, *Colder Than Hell,* 242.

244 "must have had a crystal ball": Videotaped oral history with Lee, recorded on August 24, 2000, Marine Corps Historical Center.

244 "too cold for government work": Davis, *Story of Ray Davis,* 114.

244 "seductive, mindless mist": Leckie, *March to Glory,* 137.

245 "The night was against them": Owen, *Colder Than Hell,* 235.

246 "No shoot, no shoot!": Ibid., 236.

246 "a chain gang of zombies": Russ, *Breakout,* 295.

246 "Exhaustion was telling on us": Owen, *Colder Than Hell,* 233.

246 "like pillars of lead": Russ, *Breakout,* 300.

34 THIS PLACE OF SUFFERING

247 "all seemed relatively well": Davis, *Story of Ray Davis,* 114.

248 "Not going any farther": Russ, *Breakout,* 299.

248 "The spirit had gone out": Ibid.

249 "pretty trigger-happy": Wilson, *Retreat, Hell!* 206.

249 "a patrol to guide you in": Russ, *Breakout,* 302.

250 "quiet oaths of disbelief": Owen, *Colder Than Hell,* 237.

250 "frozen in spasms of pain": Ibid., 236.

251 "a Hollywood battle set": Berry, *Hey Mac, Where Ya Been?* 158.

251 "speed that seemed to mock": Leckie, *March to Glory,* 139.

251 "Men hobbled about": Owen, *Colder Than Hell,* 237.

251 "saviors would gaze upon the saved": Leckie, *March to Glory,* 140.

251 "It was exhilarating": Videotaped oral history with Lee, recorded on August 24, 2000, Marine Corps Historical Center.

251 "We never claimed that we saved Fox": Ibid.

253 "We were stunned": Russ, *Breakout,* 303.

253 "A single bullet pierced": Hammel, *Chosin,* 271.

253 "docile little fellows": Russ, *Breakout,* 303.

254 "the usual boring Communist propaganda": Ibid., 304.

254 "threat to the motherland": Videotaped oral history with Lee.

254 "It's out of our control": Russ, *Breakout,* 304.

BOOK FIVE: TO THE SEA

35 ATTACKING IN A DIFFERENT DIRECTION

258   starting to look "like LaGuardia": Shisler, *For Country and Corps*, 210.

259   "You constantly felt the presence": Author interview with Dr. James Stewart, in Vero Beach, FL, on March 27, 2015.

260   "They had the dazed air": Higgins, *War in Korea*, 182.

260   "drunk with fatigue": Ibid., 190.

260   "snap like a pretzel": Author interview with David Douglas Duncan in Provence, France, August 29, 2015.

261   Peking Radio announced: Higgins, *War in Korea*, 181.

261   "If you have a father": Winchell is quoted in Shisler, *For Country and Corps*, 228.

261   "Retreat, *hell!*": Smith's remarks are detailed in Leckie, *March to Glory*, 169, and Shisler, *For Country and Corps*, 217.

262   "Luckily, no control cables": Author interview with Robert Himmerich y Valencia in Lone Butte, NM, on January 27, 2018.

262   "It was like a shooting gallery": Author interview with Robert Harbula, in Tampa, FL, February 29, 2016.

262   "Tootsie Rolls": See "How Tootsie Rolls Accidentally Saved Marines During War," at the Marine Corps Community Services website (www .usmc-mccs.org). At reunions of the Chosin Few, a national organization of Chosin Reservoir veterans, the Chicago-based Tootsie Roll Industries has long been a sponsor, offering prodigious supplies of the candy that sustained the men all those years ago.

264   "Our parkas were all stained": Owen, *Colder Than Hell*, 245.

264   "never *did* stand a chance": This anecdote is found in Shisler, *For Country and Corps*, 223.

264   *Here's health to you:* The full lyrics of "The Marines' Hymn" are available at the official website of the United States Marine Corps (www .marineband.marines.mil).

265   "those magnificent bastards": See Frank Kerr, "At the Reservoir: Through the Eyes of a Combat Photographer," *Leatherneck,* December 1990.

36 IN THE DAY OF TROUBLE

266   In Korea, he'd found both: My account of Ed Reeves's harrowing survival story east of Chosin is drawn primarily from his published war memoir, *Beautiful Feet & Real Peace.* Unless otherwise noted, all details, including remembered dialogue, come from his own recollections in this book. I also relied upon a lengthy oral history Reeves videotaped for the Veterans History Project in 2003, available at the website of the Library of Congress's American Folklife Center (https://www.loc.gov/folklife). See also Patricia Brennan, "The 'Forgotten' Conflict," *Washington Post,* May 28, 2000.

267   A nightmare unfolded: The napalm incident is described in Hammel, *Chosin,* 231, and in Russ, *Breakout,* 275. My description here is also drawn from my interview with Army veteran Robert Ayala, an eyewitness to the incident, and a survivor of the east-of-Chosin ordeal. I interviewed Ayala in Dallas on June 24, 2016.

267   "It was so hot out there": Author interview with Robert Ayala.

268 "The smell was awful": Author interview with Harrison Ager in San Diego, CA, on August 20, 2016.

269 Colonel Allan MacLean, was gone: MacLean's wounding and disappearance is depicted in Hammel, *Chosin*, 196.

270 "a struggling organism": Appleman, *East of Chosin*, 306.

270 "It was heartrending": Author interview with Sam Folsom, in Santa Monica, CA, on January 24, 2016.

271 "like Odysseus escaping": Russ, *Breakout*, 279.

271 he shot them both: This incident is described in Appleman, *East of Chosin*, 240, and Russ, *Breakout*, 279.

271 Faith was mortally wounded: Faith's death is detailed in Hammel, *Chosin*, 239; Simmons, *Frozen Chosin*, 77; and Appleman, *East of Chosin*, 276–77.

271 "rather like El Cid": Simmons, *Frozen Chosin*, 77.

37   I'LL GET YOU A GODDAMN BRIDGE

275 "My personal, private reaction": Shisler, *For Country and Corps*, 224.

276 "damned serious situation": Russ, *Breakout*, 357.

277 "When a cat is at the rat hole": Hopkins, *One Bugle, No Drums*, 143.

277 "most difficult defile": Russ, *Breakout*, 394.

278 "Never heard of it, General": Ibid., 358.

279 "Dammit sir!": Leckie, *March to Glory*, 180.

279 "On such slender threads": Brady, *Marines of Autumn*, 223.

38   BLOOD ON THE ICE

281 Ed Reeves didn't understand: As with the previous chapter, the Reeves narrative comes mainly from his war memoir, *Beautiful Feet & Real Peace*, and from the oral history Reeves videotaped for the Library of Congress's American Folklife Center.

286 Private First Class Ralph Milton: This part of Reeves's narrative is drawn substantially from my interview with Chosin veteran Ralph Milton in Twin Falls, ID, in July 2015.

286 "We developed a technique": Author interview with Linus Chism on August 21, 2016, in San Diego, CA.

39   TAKING DEPARTURE

289 On the afternoon of December 4: My narrative of aviators Jesse Brown and Tom Hudner is drawn primarily from two excellent written sources: *Devotion*, by Adam Makos, and *The Flight of Jesse Leroy Brown*, by Theodore Taylor. I also relied on my own conversation with Tom Hudner and his family at his home in Concord, MA, on December 13, 2015. Another helpful source was *Such Men As These*, by David Sears.

290 "Make mine a gin and tonic": Makos, *Devotion*, 160.

291 "the cabin shook": Ibid., 30.

291 "cain't see to cain't see": Taylor, *Flight of Jesse Leroy Brown*, 27.

291 "If Negroes can't ride": Makos, *Devotion*, 28.

292 "movie star handsome": Taylor, *Flight of Jesse Leroy Brown*, 200.

293 a painting he'd seen in his youth: Makos, *Devotion*, 334.

294 "looks like you're bleeding fuel!": Ibid., 9.

294 "You've got a streamer": Ibid.

294   "I have to put it down": Sears, *Such Men As These*, 119.
295   "Taking departure": Makos, *Devotion*, 337.
295   *My own dear sweet Angel:* Brown's entire letter to his wife is found in Taylor, *The Flight of Jesse Leroy Brown*, 268–71.

40   THE BRIDGE OF LONG LIFE
297   Later that same week: All the details in this chapter are drawn from my interviews with Dr. Lee Bae-suk at his home in Cincinnati.

41   DOWN TO EARTH
300   "destroy the plane": Makos, *Devotion*, 340.
301   "I'm going in": Sears, *Such Men As These*, 121.
302   "curled his fingers into claws": Makos, *Devotion*, 344.
302   "I'm pinned in here": Sears, *Such Men As These*, 122.
302   "We gotta do something": Ibid.
303   "Jesse's pinned inside": Ibid., 276.
303   "Tell Daisy how much I love her": Sears, *Such Men As These*, 123.
304   "That Jesse over there?": Makos, *Devotion*, 353.
304   "Hey Mississippi": Ibid., 354.
304   Brown . . . "half smiled": Jacobs, *Korea's Heroes*, 75.
304   "He ain't moving": Makos, *Devotion*, 354.
305   *If only my heart could talk:* Jesse Brown to Daisy Brown, December 3, 1950, reprinted in Taylor, *Flight of Jesse Leroy Brown*, 268–71.

42   THE MOST HARROWING HOUR
308   "we had to get them out of there": Russ, *Breakout*, 414.
309   "They were sick": Leckie, *March to Glory*, 205.
309   "the most harrowing hour": Russ, *Breakout*, 414.

43   THE CROSSING
310   "Seeing that star": Author interview with Manert Kennedy in Monte Vista, CO, on April 14, 2016.
312   "high-wire acrobatics": Brady, *Marines of Autumn*, 223.
313   technical sergeant named Wilfred Prosser: The story of Prosser's assuming command of the stricken bulldozer is detailed in Leckie, *March to Glory*, 213.
314   "Just for the hell of it": Russ, *Breakout*, 416.
314   "There seemed to be a glow": Partridge is quoted in Montross and Canzona, *Chosin Reservoir Campaign*, 323.
315   "Those poor folks": Author interview with Manert Kennedy in Monte Vista, CO, on April 14, 2016.

44   WE WILL SEE YOU AGAIN IN THE SOUTH
316   Lee Bae-suk found it surreal: All the details in this chapter are drawn from my interviews with Dr. Lee Bae-suk at his home in Cincinnati.

45   WE WALK IN THE HAND OF GOD
323   "Seen from the air": AP reporter Hal Boyle is quoted in Weintraub, *Christmas Far from Home*, 219.

324 *Bless 'em all!:* The chant is quoted in Leckie, *March to Glory,* 217.

324 "They had been there and back": Ibid.

324 "beat of a tragic rhythm": Duncan, *This Is War!,* 201.

324 "Give me tomorrow": Ibid.

325 "When we hit the valley": Russ, *Breakout,* 429.

325 "a sullen, brown river": Hopkins, *One Bugle, No Drums,* 200.

326 "I'd follow him to hell": Shisler, *For Country and Corps,* 232.

326 "Retreat, you say?": Russ, *Breakout,* 433.

327 "committed the unforgivable sin": Brady, *Marines of Autumn,* 264.

327 "The running fight of the Marines": *Time,* December 18, 1950.

327 "one of the greatest fighting retreats": Hersey, "Mr. President."

327 "the most efficient and courageous": Shisler, *For Country and Corps,* 238.

327 "there was a plan": La Bree, *Gentle Warrior,* 184.

327 "I knew what we had to do": Smith to Esther Smith, December 26, 1950, Box 2, General Oliver Smith Collection.

327 "I have never commanded": Shisler, *For Country and Corps,* 230.

328 "walked in the hand of God": Ibid.

328 "out of great faith can come a miracle": S. L. A. Marshall is quoted in Shisler, *For Country and Corps,* 232.

328 "To say that I am proud": Esther Smith to Smith, December 10, 1950, Box 2, General Oliver Smith Collection.

328 "My admiration for your calmness": Esther Smith to Smith, January 10, 1951, Box 2, General Oliver Smith Collection.

328 "I never felt better": Shisler, *For Country and Corps,* 240.

329 "still full of fight": La Bree, *Gentle Warrior,* 183.

329 "No more gallant men": Russ, *Breakout,* 434.

330 swirling down the drains: Ibid., 431.

330 "When I dried off": Knox, *The Korean War,* 614.

330 "They cut our boots off": Ibid., 615.

331 "May hate cease": Weintraub, *Christmas Far from Home,* 238.

331 "On an occasion like this": Shisler, *For Country and Corps,* 238.

331 "with our spirit unbroken": Smith speech, Box 29, General Oliver Smith Collection.

EPILOGUE: IN THE PANTHEON

333 "They have gone": Spurr, *Enter the Dragon,* 269.

333 "I'm not a hero": Author interview with Hector Cafferata.

335 "I wasn't planning on *biting*": Author interview with Stuart Yancey, John Yancey's son, on May 27, 2017, in St. Louis, MO.

337 "I am forever grateful": Hopkins, *One Bugle, No Drums,* 230.

SELECTED BIBLIOGRAPHY

ARCHIVES

Douglas MacArthur Memorial Library and Archives, Norfolk, VA.
George C. Marshall Research Library, Virginia Military Institute, Lexington, VA.
Harry S. Truman Presidential Library, Independence, MO.
Marine Corps History Division, Archives Branch, Quantico, VA.
National Archives, Washington, DC, and College Park, MD.
U.S. Army Heritage and Education Center, Carlisle, PA.

MUSEUMS AND MONUMENTS

Chosin Few Battle Monument, Quantico, VA.
George C. Marshall Museum, Virginia Military Institute, Lexington, VA.
Harry S. Truman Museum and Home, Independence, MO.
Korean War National Museum, Springfield, IL.
Korean War Veterans Memorial, Washington, DC.
MacArthur Museum of Arkansas Military History, Little Rock, AR.
Marine Corps Mechanized Museum, Camp Pendleton, CA.
National Museum of the Marine Corps, Triangle, VA.
National Museum of the United States Air Force, Wright-Patterson Air Force Base, Dayton, OH.
Veterans Museum at Balboa Park, San Diego, CA.
VMI Museum, Virginia Military Institute, Lexington, VA.

SPECIAL COLLECTIONS, PERSONAL PAPERS, AND ORAL HISTORIES

Almond, General Edward. Collection. Army Military History Institute, Carlisle, PA.
Battle of Chosin Archive. Full interview transcripts collected for the PBS program "The Battle of Chosin," *American Experience*, WGBH, Boston.
Duncan, David Douglas. Archive. Harry Ransom Center, University of Texas, Austin.
Faith, Colonel Don Carlos. Personal Papers. Baton Rouge, LA.
Higgins, Marguerite. Papers. Special Collections Research Center, Syracuse University, Syracuse, NY.
Polk, Lt. Col. James H. Korean War Letters. Private collection. Santa Fe, NM.

Smith, General Oliver Prince. Collection. Marine Corps History Division, Archives Branch, Quantico, VA.
Yancey Collection. MacArthur Museum of Arkansas Military History, Little Rock, AR.
Yancey, John. Personal Papers. Courtesy of the Yancey Family, Little Rock, AR.

SELECTED WEBSITES AND DATABASES

Chosin Reservoir: Lest We Forget. http://www.chosinreservoir.com
Congressional Medal of Honor Society. http://www.cmohs.org
Korean War Educator. http://www.koreanwar-educator.org/topics/chosin
Korean War Project. https://www.koreanwar.org
United States Army in the Korean War. https://www.army.mil/koreanwar

VETERANS GROUP REUNIONS ATTENDED

Chosin Few National Reunion. San Diego, CA, August 2016.
Reunion of 41 Independent Commando, the Royal Marines. London, September 2015.
U.S. Army Chapter of the Chosin Few. Springfield, MO, April 2016.

VETERANS' PUBLICATIONS AND MILITARY PAMPHLETS

Alexander, Col. Joseph H. *Battle of the Barricades: U.S. Marines in the Recapture of Seoul.* Marines in the Korean War Commemorative Series. Washington, DC: History and Museums Division, Headquarters, U.S. Marine Corps, 2000.
The Chosin Few. Official news digest of the Chosin Few, Exclusive Fraternity of Honor.
Condon, Major General John P. *Corsairs to Panthers: U.S. Marine Aviation in Korea.* Marines in the Korean War Commemorative Series. Washington, DC: History and Museums Division, Headquarters, U.S. Marine Corps, 2002.
The Fight to Save Fox Company. Marine Corps League, Summer 2000.
The Graybeards. Official publication of the Korean War Veterans Association.
Leatherneck. Magazine of the Marines.
MacDonald, James Angus, Jr. "The Problems of U.S. Marine Corps Prisoners of War in Korea." Washington, DC: History and Museums Division, Headquarters, U.S. Marine Corps, 1988.
O'Brien, Cyril J. Simmons, Brigadier General Edwin H. *Frozen Chosin: U.S. Marines at the Changjin Reservoir.* Marines in the Korean War Commemorative Series. Washington, DC: History and Museums Division, Headquarters, U.S. Marine Corps, 2002.
The Reservoir. Newsletter of the Chosin Few, New England chapter.
Simmons, Brigadier General Edwin H. *Over the Seawall: U.S. Marines at Inchon.* Marines in the Korean War Commemorative Series. Washington, DC: History and Museums Division, Headquarters, U.S. Marine Corps, 2000.
"Story About God's Miracle During the Korean War: The Heungnam Evacuation Operation, December of 1950." Heungnam Evacuation Operation Commemoration Committee. May 27, 2005.

Thomas, Lt. Col. Peter. *41 Independent Commando RM, Korea 1950–1952*. Royal Marines Historical Society, Special Publication No. 8, 1985.

DOCUMENTARIES

Battle for Korea. Produced by Malin Film and Television in association with PBS, 2001.

The Battle of Chosin. Directed by Randall MacLowry and produced by the Film Posse for *American Experience*, WGBH, Boston, 2016.

Chosin. Directed by Brian Iglesias. Produced by Veterans Inc. in association with Post Factory NY Films, 2010.

Korea: The Forgotten War. Produced by A&E Television Networks for the History Channel, 1987.

The Korean War. 60th Anniversary Commemorative Documentary Collection. Mill Creek Entertainment, 2013.

The Korean War: Fire and Ice. Lou Reda Productions, 2010.

Task Force Faith: The Story of the 31st Regimental Combat Team. Produced and directed by Julie Precious, 2013.

Uncommon Courage: Breakout from Chosin. Produced by KPI Television for the Smithsonian Channel, 2010.

SELECTED MAGAZINE ARTICLES

Bell, James. "The Brave Men of No Name Ridge." *Life*, August 28, 1950.

Chandler, Lt. Col. James B. "A Letter Home." *Marine Corps Gazette*, March 1988.

Geer, Colonel Andrew. "Breakout!" *American Weekly*, December 1, 1957.

"General O. P. Smith, U.S.M.C.: By Land, Sea, and Air." *Time*, September 25, 1950.

Hersey, John. "Mr. President." *New Yorker*, May 5, 1951.

Higgins, Marguerite. "The Bloody Trail Back." *Saturday Evening Post*, January 27, 1951.

Jaskilka, General Samuel. "Then and Now: Easy Alley." *Leatherneck*, November 1990.

Keene, R. R. "The Division Breaks Out." *Leatherneck*, December 1990.

Kerr, Frank. "At the Reservoir: Through the Eyes of a Combat Photographer." *Leatherneck*, December 1990.

Morgan, Martin K. A. "When the Deer Come Running." *American Rifleman*, February 2016.

Moskin, J. Robert. "Chosin." *American Heritage*, November 2000.

"Now It Can Be Told: What Happened in Korea When Chinese Marched In. Interview with Lieut. Gen. E. M. Almond." *U.S. News & World Report*, February 13, 1953.

O'Neill, Tom. "Dangerous Divide." *National Geographic*, July 2003.

Smith, Oliver Prince. "Looking Back at Chosin." *Marine Corps Gazette*, November 2000.

"There Was a Christmas in Korea." *Life*, December 25, 1950.

Wallace, James. "Bloody Chosin: The Blind Lead the Brave." *U.S. News & World Report*, June 25, 1990.

BOOKS

Acheson, Dean. *The Korean War.* New York: W. W. Norton, 1969.

———. *Present at the Creation: My Years in the State Department.* New York: W. W. Norton, 1969.

Appleman, Roy E. *East of Chosin: Entrapment and Breakout in Korea, 1950.* College Station: Texas A&M University Press, 1987.

———. *Escaping the Trap: The US Army X Corps in Northeast Korea, 1950.* College Station: Texas A&M University Press, 1990.

———. *United States Army in the Korean War,* vol. 2: *South to the Naktong, North to the Yalu.* Washington, DC: Office of the Chief of Military History, Department of the Army, 1961.

Atkinson, Rick. *The Day of Battle: The War in Sicily and Italy, 1943–1944.* New York: Henry Holt, 2008.

Barenblatt, Daniel. *A Plague Upon Humanity: The Hidden History of Japan's Biological Warfare Program.* New York: HarperPerennial, 2005.

Bartlett, Merrill L., and Jack Sweetman. *Leathernecks: An Illustrated History of the United States Marine Corps.* Annapolis, MD: Naval Institute Press, 2008.

———. *The U.S. Marine Corps: An Illustrated History.* Annapolis, MD: Naval Institute Press, 2001.

Beisner, Robert L. *Dean Acheson: A Life in the Cold War.* New York: Oxford University Press, 2006.

Bernstein, Richard. *China 1945: Mao's Revolution and America's Fateful Choice.* New York: Vintage, 2014.

Berry, Henry. *Hey, Mac, Where Ya Been? Living Memories of the U.S. Marines in the Korean War.* New York: St. Martin's, 1988.

Blair, Clay. *The Forgotten War: America in Korea 1950–1953.* New York: Anchor, 1987.

Brady, James. *The Marines of Autumn: A Novel of the Korean War.* New York: Thomas Dunne, 2000.

Brands, H. W. *The General vs. the President: MacArthur and Truman at the Brink of Nuclear War.* New York: Doubleday, 2016.

Cha, Victor. *The Impossible State: North Korea, Past and Future.* New York: Harper Collins, 2012.

Chang, Jung, and Jon Halliday. *Mao: The Unknown Story.* New York: Alfred A. Knopf, 2005.

Chapman, Franklin "Jack." *Cherokee Warrior.* A copyrighted memoir published on the Korean War Educator website: http://www.koreanwar-educator.org /memoirs/chapman_franklin_jack. December 2014.

Choe, Sang-Hung, and Christopher Torchia. *How Koreans Talk: A Collection of Expressions.* Seoul: EunHaeng Namu, 2002.

Chung, Donald K., M.D. *The Three Day Promise: A Korean Soldier's Memoir.* Tallahassee, FL: Father & Son, 1989.

Clark, Commander Eugene Franklin. *The Secrets of Inchon: The Untold Story of the Most Daring Covert Mission of the Korean War.* New York: Berkley, 2002.

Cleaver, Thomas McKelvey. *The Frozen Chosen: The 1st Marine Division and the Battle of the Chosin Reservoir.* Oxford, UK: Osprey, 2016.

Coggins, Jack. *Soldiers and Warriors: An Illustrated History.* Mineola, NY: Dover Publications, 1966.

Connor, John W. *Let Slip the Dogs of War: A Memoir of the GHQ 1st Raider Company (8245th Army Unit) a.k.a. Special Operations Company, Korea 1950–51.* Bennington, VT: Merriam Press, 2008.

Cook, Fred J. *The Nightmare Decade: The Life and Times of Senator Joe McCarthy.* New York: Random House, 1971.

Coye, Molly Joel, Jon Livingston, and Jean Highland, eds. *China: Yesterday and Today.* New York: Bantam, 1975.

Crumbie, SSgt. Watson A., Jr. *My Time As a Marine, 1943–1952.* Privately published.

Cumings, Bruce. *The Korean War.* New York: Modern Library, 2010.

Davis, Burke. *Marine! The Life of Chesty Puller.* New York: Bantam, 1962.

Davis, Ray. *The Story of Ray Davis.* Varina, NC: Research Triangle Publishing, 1995.

Domes, Jürgen. *Peng Te-huai: The Man and the Image.* Stanford, CA: Stanford University Press, 1985.

Donovan, Robert J. *The Assassins.* London: Elek Books, 1956.

———. *Conflict and Crisis: The Presidency of Harry S. Truman, 1945–1948.* New York: W. W. Norton, 1977.

———. *Tumultuous Years: The Presidency of Harry S. Truman, 1949–1953.* New York: W. W. Norton, 1982.

Dower, John W. *Embracing Defeat: Japan in the Wake of World War II.* New York: W. W. Norton, 1999.

Drury, Bob, and Tom Clavin. *The Last Stand of Fox Company: A True Story of U.S. Marines in Combat.* New York: Grove, 2009.

Duncan, David Douglas. *My 20th Century.* New York: Arcade, 2014.

———. *This Is War! A Photo-Narrative of the Korean War.* Boston: Little, Brown, 1951.

Fehrenbach, T. R. *This Kind of War: The Classic Korean War History.* Washington, DC: Brassey's, 1963.

Flanagan, Richard. *The Narrow Road to the Deep North: A Novel.* New York: Alfred A. Knopf, 2013.

Geer, Andrew. *The New Breed: The Story of the U.S. Marines in Korea.* New York: Harper & Brothers, 1952.

Gilbert, Bill. *Ship of Miracles: 14,000 Lives and One Miraculous Voyage.* Chicago: Triumph, 2000.

Goncharov, Sergei N., John W. Lewis, and Xue Litai. *Uncertain Partners: Stalin, Mao, and the Korean War.* Stanford, CA: Stanford University Press, 1993.

Gordon, Andrew. *A Modern History of Japan: From Tokugawa Times to the Present.* New York: Oxford University Press, 2009.

Goulden, Joseph C. *Korea: The Untold Story of the War.* New York: Times Books, 1982.

Gray, John Edward. *Called to Honor: Memoirs of a Three-War Veteran.* Asheville, NC: R. Brent, 2006.

Griffin, W. E. B. *Retreat, Hell! A Novel of the Corps.* New York: G. P. Putnam's Sons, 2004.

Gunther, John. *The Riddle of MacArthur: Japan, Korea and the Far East.* New York: Harper & Brothers, 1950.

Haig, Alexander M., with Charles McCarry. *Inner Circles: How America Changed the World.* New York: Warner, 1992.

Halberstam, David. *The Coldest Winter: America and the Korean War*. New York: Hyperion, 2007.

Hammel, Eric. *Chosin: Heroic Ordeal of the Korean War*. Minneapolis: Zenith Press, 1981.

Harden, Blaine. *The Great Leader and the Fighter Pilot: A True Story About the Birth of Tyranny in North Korea*. New York: Penguin, 2015.

——. *King of Spies: The Dark Reign of America's Spymaster in Korea*. New York: Viking, 2017.

Harper, Merrill. *Chosin Reservoir: As I Remember Koto-ri Pass, North Korea, December 1950*. Bloomington, IN: iUniverse, 2012.

Hastings, Max. *The Korean War*. New York: Simon & Schuster Paperbacks, 1987.

Hayhurst, Fred. *Green Berets in Korea: The Story of 41 Independent Commando Royal Marines*. Cambridge, UK: Vanguard Press, 2001.

Haynes, Major Justin M. *Intelligence Failure in Korea: Major General Charles A. Willoughby's Role in the United Nations Command's Defeat in November, 1950*. New York: Pickle Partners Publishing, 2015.

Heinl, Robert Debs. *Soldiers of the Sea: The United States Marine Corps, 1775–1962*. Baltimore: Nautical & Aviation Publishing Company of America, 1991.

——. *Victory at High Tide: The Inchon-Seoul Campaign*. Baltimore: Nautical & Aviation Publishing Company of America, 1979.

Hersey, John. *Aspects of the Presidency*. New Haven, CT: Ticknor & Fields, 1980.

Hicks, George. *The Comfort Women: Japan's Brutal Regime of Enforced Prostitution in the Second World War*. New York: W. W. Norton, 1997.

Higgins, Marguerite. *War in Korea: The Report of a Woman Combat Correspondent*. Garden City, NY: Doubleday, 1951.

Hoffman, Jon T. *Chesty: The Story of Lieutenant General Lewis B. Puller, USMC*. New York: Random House, 2001.

Hopkins, William B. *One Bugle, No Drums: The Marines at Chosin Reservoir*. Chapel Hill, NC: Algonquin Books of Chapel Hill, 1986.

Hunt, George P. *The Story of the U.S. Marines*. New York: Random House, 1951.

Hunter, Stephen, and John Bainbridge. *American Gunfight: The Plot to Kill Harry Truman—and the Shoot-out That Stopped It*. New York: Simon & Schuster, 2005.

Jacobs, Bruce. *Korea's Heroes: The Medal of Honor Story*. New York: Berkley, 1961.

Jacobs, Ralph. *Cord-Wood: A Collection of Korean War Poems*. Self-published, 2004.

Jager, Sheila Miyoshi. *Brothers at War: The Unending Conflict in Korea*. New York: W. W. Norton, 2013.

James, D. Clayton. *The Years of MacArthur*, vol. 3: *Triumph & Disaster, 1945–1964*. Boston: Houghton Mifflin, 1985.

Janca, Richard A. *The Relic: A Marine's Memoir of the Korean War*. Palm Beach Gardens, FL: Marquesas, 2012.

Jeon, Kyung-Ae. *The Dandelion Ranch*. Seoul: Korea Herald Inc., 2000.

Jian, Chen. *China's Road to the Korean War: The Making of the Sino-American Confrontation*. New York: Columbia University Press, 1994.

——. *Mao's China and the Cold War*. Chapel Hill: University of North Carolina Press, 2001.

Jin, Ha. *War Trash*. New York: Pantheon, 2004.

Johnson, Adam. *The Orphan Master's Son: A Novel*. New York: Random House, 2012.

Kang, Hildi. *Under the Black Umbrella: Voices from Colonial Korea, 1910–1945.* Ithaca, NY: Cornell University Press, 2001.

Khlevniuk, Oleg V. *Stalin: New Biography of a Dictator.* New Haven, CT: Yale University Press, 2015.

Kim Il Sung. *The Selected Works of Kim Il Sung.* New York: Prism Key Press, 2011.

Klara, Robert. *The Hidden White House: Harry Truman and the Reconstruction of America's Most Famous Residence.* New York: St. Martin's Press, 2013.

Knox, Donald. *The Korean War: Pusan to Chosin.* San Diego: Harcourt Brace Jovanovich, 1985.

Krulak, Victor H. *First to Fight: An Inside View of the U.S. Marine Corps.* Annapolis, MD: Naval Institute Press, 1984.

La Bree, Clifton. *The Gentle Warrior: General Oliver Prince Smith, USMC.* Kent, OH: Kent State University Press, 2001.

Langley, Michael. *Inchon Landing: MacArthur's Last Triumph.* New York: Times Books, 1979.

Leckie, Robert. *March to Glory.* New York: Simon & Schuster, 1960.

Lewis, Jack. *Chosen Tales of Chosin.* North Hollywood, CA: Challenge, 1964.

Li, Xiaobing. *A History of the Modern Chinese Army.* Lexington: University Press of Kentucky, 2007.

———. *They Came in Waves: The Chinese Attack at Chosin.* Unpublished manuscript, December 2017, submitted to University of Oklahoma Press.

Li, Xiaobing, Allan R. Millett, and Bin Yu, eds. *Mao's Generals Remember Korea.* Lawrence: University Press of Kansas, 2001.

Li Zhisui. *The Private Life of Chairman Mao: The Memoirs of Mao's Personal Physician.* New York: Random House, 1994.

MacArthur, Douglas. *Reminiscences.* New York: McGraw-Hill, 1964.

Maffioli, Len, and Bruce H. Norton. *Grown Gray in War: From Iwo Jima to the Chosin Reservoir to the Tet Offensive, the Autobiography of a True Marine Hero.* San Diego: Quadrant, 1997.

Makos, Adam. *Devotion: An Epic Story of Heroism, Friendship, and Sacrifice.* New York: Ballantine, 2015.

Manchester, William. *American Caesar: Douglas MacArthur 1880–1964.* Boston: Little, Brown, 1978.

Mao Tse-tung. *On Guerrilla Warfare.* Westport, CT: Praeger, 1961.

———. *Quotations from Chairman Mao Tse-tung.* Peking: Foreign Languages Press, 1972.

———. *Talks at the Yenan Forum on Literature and Art.* Peking: Foreign Languages Press, 1967.

Marshall, S. L. A. *The River and the Gauntlet.* Chicago: Time-Life Books, 1962.

Martin, Bradley K. *Under the Loving Care of the Fatherly Leader: North Korea and the Kim Dynasty.* New York: Thomas Dunne, 2004.

May, Antoinette. *Witness to War.* New York: Penguin, 1983.

McCloskey, Pete. *A Year in a Marine Rifle Company: Korea, 1950–51.* Sunnyvale, CA: Patsons Press, 2013.

———. *The Taking of Hill 610.* Woodside, CA: Eaglet, 1992.

McCullough, David. *Truman.* New York: Simon & Schuster, 1992.

McGovern, James. *To the Yalu.* New York: William Morrow, 1972.

McLellan, David S., and David C. Acheson, eds. *Among Friends: Personal Letters of Dean Acheson.* New York: Dodd, Mead, 1980.

Mill, John Stuart. *Principles of Political Economy.* New York: Prometheus, 2014.

Miller, Merle. *An Oral Biography of Harry S. Truman.* New York: Berkley, 1973.

Montross, Lynn, and Nicholas A. Canzona. *U. S. Marine Operations in Korea, 1950–1953,* vol. 3: *The Chosin Reservoir Campaign.* Austin, TX: R. J. Speights, 1992.

Mowat, Farley. *Eastern Passage.* Toronto: Emblem Editions, 2011.

Murray, Zona Gayle. *Highpockets: The Man, the Marine, the Legend.* Self-published, 2009.

Nicholson, Jim. *George-3-7th Marines: A Brief Glimpse through Time of a Group of Young Marines.* Dallas: Brown Books, 2015.

O'Donnell, Patrick K. *Give Me Tomorrow: The Korean War's Greatest Untold Story—The Epic Stand of the Marines of George Company.* Cambridge, MA: Da Capo Press, 2010.

Olmstead, Robert. *The Coldest Night.* Chapel Hill, NC: Algonquin Books of Chapel Hill, 2012.

Oshinsky, David M. *A Conspiracy So Immense: The World of Joe McCarthy.* New York: Free Press, 1983.

Oskins, James C., and Michael H. Maggelet. *Broken Arrow: The Declassified History of U.S. Nuclear Weapons Accidents.* Raleigh, NC: Lulu.com, 2008.

Owen, Joseph R. *Colder Than Hell: A Marine Rifle Company at Chosin Reservoir.* New York: Ivy, 1996.

Paik Sun Yup. *From Pusan to Panmunjom.* Dulles, VA: Potomac, 1992.

Pearlman, Michael D. *Truman and MacArthur: Policy, Politics, and the Hunger for Honor and Renown.* Bloomington: Indiana University Press, 2008.

Peng Dehuai. *Memoirs of a Chinese Marshal: The Autobiographical Notes of Peng Dehuai (1898–1974).* Honolulu: University Press of the Pacific, 2005.

Perry, Mark. *The Most Dangerous Man in America: The Making of Douglas MacArthur.* New York: Basic Books, 2014.

Peters, Richard, and Xiaobing Li. *Voices form the Korean War: Personal Stories of American, Korean, and Chinese Soldiers.* Lexington: University Press of Kentucky, 2004.

Polk, James H. *World War II Letters and Notes of Colonel James H. Polk 1944–1945.* Edited by James H. Polk III. Oakland, OR: Red Anvil Press, 2005.

Puller, Lewis B. *Fortunate Son: The Autobiography of Lewis B. Puller Jr.* New York: Grove Weidenfeld, 1991.

Quigley, Bill. *Passage Through a Hell of Fire and Ice: Korea . . . the First Five Months, a Marine Epic.* New York: Page Publishing, 2015.

Radzinsky, Edvard. *Stalin: The First In-Depth Biography Based on Explosive New Documents from Russia's Secret Archives.* New York: Doubleday, 1996.

Ragland, Nick, and Joe Rouse. *Puller's Runner: A Work of Historical Fiction About Lieutenant General Lewis B. "Chesty" Puller.* Lanham, MD: Hamilton, 2009.

Ragland, Nick, and Tom Schwettman. *The Savior: General Oliver Prince Smith.* Wilmington, OH: Orange Frazer Press, 2016.

Reeves, Hubert Edward. *Beautiful Feet & Real Peace.* Prescott, AZ: Melcher Printing, 1997.

Richardson, William, and Kevin Maurer. *Valleys of Death: A Memoir of the Korean War.* New York: Berkley, 2010.

Ricks, Thomas E. *The Generals: American Military Command from World War II to Today.* New York: Penguin Press, 2012.

Roe, Patrick C. *The Dragon Strikes: China and the Korean War, June–December 1950*. Novato, CA: Presidio Press, 2000.

Rottman, Gordon L. *Inchon 1950: The Last Great Amphibious Assault*. Oxford, UK: Osprey, 2006.

Rovere, Richard. *Senator Joe McCarthy*. Berkeley: University of California Press, 1996.

Rowny, Edward L. *An American Soldier's Saga of the Korean War*. Washington, DC: Self-published, 2013.

Russ, Martin. *Breakout: The Chosin Reservoir Campaign, Korea 1950*. New York: Penguin, 1999.

————. *The Last Parallel: A Marine's War Journal*. New York: Fromm International, 1957.

Salmon, Andrew. *Scorched Earth, Black Snow: Britain and Australia in the Korean War, 1950*. London: Aurum Press, 2011.

Salter, James. *Burning the Days*. London: Picador, 1997.

Sears, David. *Such Men As These: The Story of the Navy Pilots Who Flew the Deadly Skies Over Korea*. New York: Da Capo Press, 2010.

Septer, Dirk. *Lost Nuke: The Last Flight of Bomber 075*. Victoria, BC: Heritage House, 2016.

Shaara, Jeff. *The Frozen Hours*. New York: Ballantine, 2017.

Shisler, Gail B. *For Country and Corps: The Life of General Oliver P. Smith*. Annapolis, MD: Naval Institute Press, 2009.

Shoemaker, Robert C. *A Surgeon Remembers: Korea 1950–1951 and the Marines*. Victoria, BC: Trafford, 2005.

Sloan, Bill. *The Darkest Summer: Pusan and Inchon 1950: The Battles That Saved South Korea—and the Marines—from Extinction*. New York: Simon & Schuster, 2009.

Smith, Larry. *Beyond Glory: Medal of Honor Heroes in Their Own Words: Extraordinary Stories of Courage from World War II to Vietnam*. New York: W. W. Norton, 2003.

Snow, Edgar. *Red Star Over China*. New York: Grove Press, 1938.

Spanier, John W. *The Truman-MacArthur Controversy and the Korean War*. New York: W. W. Norton, 1965.

Spence, Jonathan. *Mao Zedong*. New York: Penguin, 1999.

————. *The Search for Modern China*. New York: W. W. Norton, 1990.

Spurr, Russell. *Enter the Dragon: China's Undeclared War Against the U.S. in Korea, 1950–51*. New York: Newmarket Press, 1988.

Stanton, Shelby L. *America's Tenth Legion: X Corps in Korea, 1950*. Novato, CA: Presidio Press, 1989.

Stewart, James H., M.D. *Notes of a Korean War Surgeon*. Privately published.

Stone I. F. *The Hidden History of the Korean War, 1950–1951: A Nonconformist History of Our Times*. Boston: Little, Brown, 1952.

Suh, Dae-Sook. *Kim Il Sung: The North Korean Leader*. New York: Columbia University Press, 1988.

Sun Tzu. *The Art of War*. Foreword by James Clavell. London: Hodder and Stoughton, 1981.

Taylor, Theodore. *The Flight of Jesse Leroy Brown*. Annapolis, MD: Naval Institute Press, 1998.

Thompson, Earl. *Tattoo*. New York: New American Library, 1974.

Thompson, Reginald. *Cry Korea: A Korean War Notebook*. London: Reportage Press, 2009.

Thornton, Richard C. *Odd Man Out: Truman, Stalin, Mao, and the Origins of the Korean War*. Dulles, VA: Brassey's, 2000.

Toland, John. *In Mortal Combat: Korea, 1950–1953*. New York: William Morrow, 1991.

Truman, Harry S. *Memoirs by Harry S. Truman: Year of Decisions*. Garden City, NY: Doubleday, 1955.

———. *Memoirs of Harry S. Truman: 1946–52, Years of Trial and Hope*. New York: Smithmark, 1955.

Truman, Margaret. *Harry S. Truman*. New York: William Morrow, 1972.

Tsegeletos, George H. *As I Recall: A Marine's Personal Story*. Bloomington, IN: 1st Books Library, 2003.

VandeLinde, Bob L. *Korea: Why Were We There? What Were We Fighting For?* Lynchburg, VA: Warwick House, 2012.

Wainstock, Dennis. *Truman, MacArthur, and the Korean War*. New York: Enigma, 1999.

Weintraub, Stanley. *A Christmas Far from Home: An Epic Tale of Courage and Survival During the Korean War*. Boston: Da Capo Press, 2014.

———. *MacArthur's War: Korea and the Undoing of an American Hero*. New York: Free Press, 2000.

Whiting, Allen S. *China Crosses the Yalu: The Decision to Enter the Korean War*. Stanford, CA: Stanford University Press, 1960.

Whitney, Courtney. *MacArthur: His Rendezvous with History*. New York: Alfred A. Knopf, 1955.

Willoughby, Charles A., and John Chamberlain. *MacArthur: 1941–1951*. New York: McGraw-Hill, 1954.

Wilson, Arthur W. *Korean Vignettes: Faces of War*. Portland, OR: Artwork Publications, 1996.

———, ed. *Red Dragon: "The Second Round," Faces of War II*. Portland, OR: Artwork Publications, 2003.

Wilson, Jim. *Retreat, Hell! We're Just Attacking in Another Direction*. New York: William Morrow, 1988.

Xenophon. *The Persian Expedition*. Translated by Rex Warner. London: Penguin, 1972.

Zhang, Shu Guang. *Mao's Military Romanticism: China and the Korean War, 1950–1953*. Lawrence: University Press of Kansas, 1995.

Zhihua, Shen. *Mao, Stalin and the Korean War: Trilateral Communist Relations in the 1950s*. Translated by Neil Silver. New York: Routledge, 2012.

INDEX

ABOUT THE AUTHOR

Hampton Sides is an award-winning editor at large for *Outside* and the author of the bestselling histories *In the Kingdom of Ice*, *Hellhound on His Trail*, *Blood and Thunder*, and *Ghost Soldiers*.